HEAT TRANSFER OF A CYLINDER IN CROSSFLOW

Experimental and Applied Heat Transfer Guide Books

A. Žukauskas, *Editor*

A. Žukauskas and *J. Žiugžda*, Heat Transfer of a Cylinder in Crossflow

IN PREPARATION

J. Stasiulevičius and *A. Skrinska*, Heat Transfer of Bundles of Finned Tubes in Crossflow

J. Vilemas, B. Česna, and *V. Survila*, Heat Transfer in Gas-Cooled Annular Channels

M. Tamonis, Radiative and Combined Heat Transfer in Channels

A. Žukauskas and *A. Šlančiauskas*, Heat Transfer in Turbulent Flow of Fluids

A. Žukauskas, V. Katinas, and *R. Ulinskas*, Vibration and Fluid Dynamics of Tube Bundles in Crossflow

A. Žukauskas and *R. Ulinskas*, Heat Transfer of Tube Bundles in Crossflow

HEAT TRANSFER OF A CYLINDER IN CROSSFLOW

A. Žukauskas

J. Žiugžda

Institute of Physical and Technical Problems of Energetics
Academy of Sciences of the Lithuanian SSR

Translated by

E. Bagdonaite

Academy of Sciences of the Lithuanian SSR

Edited by

G. F. Hewitt

AERE Harwell, U.K.

HEMISPHERE PUBLISHING CORPORATION

Washington New York London

DISTRIBUTION OUTSIDE NORTH AMERICA

SPRINGER-VERLAG

Berlin Heidelberg New York Tokyo

HEAT TRANSFER OF A CYLINDER IN CROSSFLOW

1 2 3 4 5 6 7 8 9 0 B C B C 8 9 8 7 6 5

This book was set in Press Roman by Hemisphere Publishing Corporation. The editors were Barbara A. Bodling, and Elizabeth Dugger; the production supervisor was Miriam Gonzalez; and the typesetter was A. Wayne Hutchins.
BookCrafters, Inc. was printer and binder.
Originally published by Mokslas, Vilnius as Teplootdach tsilindra v poperechnom potoke zhidkosti in the Series Teplofizika, Volume II.

Library of Congress Cataloging in Publication Data

Žukauskas, A., date
Heat transfer of a cylinder in crossflow.

(Experimental and applied heat transfer guide books)
Translation of Teplootdacha tsilindra v poperechnom potoke zhidkosti.
Bibliography: p.
Includes index.
1. Heat–Transmission. 2. Cylinders–Fluid dynamics. 3. Heat exchangers. I. Žiugžda, J. II. Title. III. Series: Experimental and applied heat transfer.
TJ260.Z4913 1985 536′.23 84-19169
ISBN 0-89116-365-4 Hemisphere Publishing Corporation

DISTRIBUTION OUTSIDE NORTH AMERICA:
ISBN 3-540-15518-X Springer-Verlag Berlin

CONTENTS

Preface ix

1 INTRODUCTION 1

2 EXPERIMENTAL TECHNIQUES 9

2.1 Experimental Equipment 10
2.2 Determination of Heat Transfer from a Cylinder 14
2.3 Determination of Fluid Dynamic Parameters 20
2.4 Notes on Experimental Procedure 25

3 ANALYTICAL TECHNIQUES 27

3.1 The Initial System of Equations 28
3.2 The Laminar Boundary Layer on the Front Part of a Cylinder 28
3.3 The Effects of Blockage Factor on the Fluid Dynamics and Heat Transfer 31
3.4 Pseudo-Laminar Fluid Dynamics and Heat Transfer at the Front Part of a Cylinder 32
3.5 Heat Transfer and Fluid Dynamics in the Transitional and Turbulent Boundary Layers on a Cylinder 36
3.6 Notes on the Implementation of the Analytical Techniques 40

4 SPECIFIC FEATURES OF FLUID DYNAMICS OF CROSSFLOW OVER A CYLINDER 45

4.1 A Comparison of Ideal and Real Fluid Dynamics over a Cylinder 45
4.2 Boundary-Layer Separation 48
4.3 Vortex Flow in the Wake 54
4.4 Shear Stress Distribution 57
4.5 Pressure Distribution 65
4.6 Velocity Distribution in the Boundary Layer 71
4.7 Hydraulic Drag 73

5 HEAT TRANSFER AT THE FRONT STAGNATION POINT ON A CYLINDER 79

5.1 Specific Features of Fluid Dynamics and Heat Transfer at the Front Stagnation Point 79
5.2 The Determining Factors of Heat Transfer at the Front Stagnation Point 85

6 LOCAL HEAT TRANSFER 97

6.1 The Circular Cylinder in the Subcritical Flow Regime 97
6.2 The Elliptic Cylinder in the Subcritical Flow Regime 108
6.3 The Circular Cylinder in the Critical Flow Regime 110
6.4 Local Heat Transfer Behavior for Flow over a Rough-Surface Cylinder 117
6.5 The Blockage Factor and Local Heat Transfer 119
6.6 Heat Transfer at the Rear Stagnation Point 126

7 AVERAGE HEAT TRANSFER 129

7.1 The Average Heat Transfer and the Fluid Physical Properties 129
7.2 The Reynolds Number and the Average Heat Transfer Coefficient 134
7.3 Summary of Correlations for the Average Heat Transfer Coefficient 138
7.4 The Effect of Free-Stream Turbulence on the Average Heat Transfer Coefficient 140
7.5 The Blockage Factor and the Average Heat Transfer Coefficient 144
7.6 The Average Heat Transfer of a Rough-Surface Cylinder 146
7.7 Average Heat Transfer Behavior of Cylinders with Noncircular Cross Section 149

8 CONCLUSION AND GENERAL REMARKS 153

8.1 The Boundary Layer and Local Heat Transfer 153
8.2 Fluid Dynamics on a Cylinder 154
8.3 Hydraulic Drag 156
8.4 Local Heat Transfer 158
8.5 The Effect of Fluid Physical Properties on Heat Transfer 160
8.6 Determination of Average Heat Transfer Coefficient 161
8.7 General Remarks 162

Appendix

1 Physical Properties of Fluids 167
2 Local Heat Transfer Data for Cylinders in Crossflow 168
3 Average Heat Transfer Data for Cylinders in Crossflow 173
4 Average Heat Transfer for Cylinders in Turbulent Flows of Air and Water 177
5 Average Heat Transfer of Two Cylinders in Series 180
6 Average Heat Transfer Data for Elliptic Cylinders 182
7 Experimental Data on the Fluid Dynamics of Circular Cylinders 183
8 Distribution of the Shear Stress and Pressure 186
9 Experimental Data on the Hydraulic Drag 190

References 193

Nomenclature 201

Index 205

PREFACE

This volume of the series presents heat transfer studies on single cylinders, performed over several years at the Institute of Physical and Technical Problems of Energetics, Academy of Sciences of the Lithuanian SSR.

Experimental data on the local and average heat transfer between a cylinder and a crossflow are given for different fluids in the ranges of Reynolds number from 1 to $2 \cdot 10^6$ and of Prandtl number from 0.7 to 1000. Information also is given on fluid dynamic parameters in various flow regimes and on the effects on heat transfer of viscosity variations caused by differences in temperature and heat flux direction.

The development of an understanding of the numerous experimental values of local and average heat transfer involves taking account of the effect of free-stream turbulence and of the related shear stress and pressure distribution. The effects of blockage of the flow channel and of tube surface roughness, among others, were studied.

The bulk of this book deals with results accumulated by the authors and interpreted by a unified technique. Descriptions of the experimental techniques are given, and characteristic test data are presented in tabular form. Some practical suggestions are made on the prediction of heat transfer and fluid dynamic parameters.

Chapter 7 was compiled by A. Žukauskas and Chapters 4, 5, and 6 by J. Žiugžda; the rest is a product of our combined efforts.

We are grateful to Prof. Dr. V. Isachenko and Prof. Dr. A. Šlančiauskas for their valuable remarks; to P. Daujotas, V. Ilgarubis, P. Vaitiekunas, G. Zdanavičius, V. Katinas, and other research engineers for their active participation in the work; and to other colleagues who assisted in bringing about the publication of this book.

A. Žukauskas
J. Žiugžda

HEAT TRANSFER OF A CYLINDER IN CROSSFLOW

CHAPTER

ONE

INTRODUCTION

Many forms of heat exchange are employed in the different branches of modern technology. Curvilinear bodies in cross flow constitute a very common group of elements in such exchangers. These may be circular cylinders (tubes), rectangular or elliptic pipes, and bodies of other geometries. Circular cylinders find perhaps the most widespread application in heat exchangers, power generators, and other thermal apparatus. The circular cylinder constitutes a classical element in boilers, in steam or gas turbines, in gas compressors, and in various aerodynamic systems. Circular cylinders or tubes of other cross section shapes are also used in buildings, as pipelines, chimneys, poles, supports, columns, etc.

A highly important field of application for a circular cylinder is in measurement—in particular, in thermal anemometry, which employs thin wires both in crossflow and at different yaw angles, as fluid-cooled sensors.

A close relationship is observed between the heat transfer behavior of a cylinder on the one hand, and the specific features of fluid dynamics on the other. The fluid dynamics of flow around a cylinder is highly complicated, due to the combined effects of the Reynolds number, the level of the free-stream turbulence, and a number of other factors.

With a cylinder in crossflow of a real fluid, a laminar boundary layer is formed on the front part as the result of the viscous forces. It is commonly accepted that, in the lower range of Reynolds number (Re), the cylinder is enveloped all around by a laminar boundary layer, which separates from its surface only at the rear stagnation point (Fig. 1*a*). An increase of Re leads to an increase in the effect of inertial forces, so that the laminar boundary layer

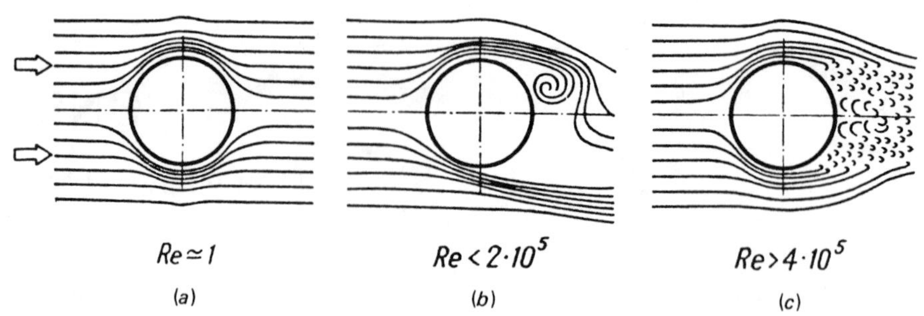

Figure 1 Patterns of fluid flow around a cylinder at different Reynolds numbers.

separates the surface at a certain distance from the rear stagnation point, and a complex vortex structure is formed in the wake (Fig. 1*b*). With a further increase of Re, the boundary layer gradually becomes turbulent, and its separation point is shifted upstream (Fig. 1*c*). This complex fluid dynamic behavior is reflected in the heat transfer between the cylinder and the fluid. The two processes have been the subject of both applied and fundamental studies for many years.

A review of early twentieth century publications on the subject shows that there was a general tendency to study the lower range of Re. The earliest studies of heat transfer were performed in air [1-3], but these were soon followed by the first attempts to study flows of liquid [4, 5]. Two major directions were covered in this stage: that of the average heat transfer from wires, so significant for thermal anemometry, and of the heat transfer from tubes in cross flow.

The second direction of research was closely connected with the progress of boiler design, and here, along with single cylinders, studies were carried out with arrays, or bundles of cylinders. In the present book we shall refer only to single cylinders. Studies of bundles of tubes were reviewed in our previous publications [6, 7].

Heat transfer studies on wires are closely related to specific problems of thermal anemometry, and as such are outside the scope of the present study. Certain aspects of the work on heat transfer from wires are useful as a source of data for the lowest range of Re. For those interested specifically in heat transfer from wires, we recommend recent studies in air [8, 9] and in liquids [10, 11].

Early studies of heat transfer from cylinders were concerned with limited numbers of parameters, and the results were presented in the form of heat transfer coefficient as a function of velocity. Such presentations were highly inconvenient in practical applications, and later research was directed toward the development of the application of the theory of similarity to the various processes. This progress in the application of theory of similarity was accompanied by numerous improvements of the experimental techniques. Application of the theory to convective heat transfer opened up the prospect of better experimental design. It also gave a common basis for comparisons of data from different sources, and for interpretation relations of versatile practical applicability. For instance, relationships of the form

$$\mathrm{Nu} = c \cdot \mathrm{Re}^{m} \cdot \mathrm{Pr}^{n} \tag{1.1}$$

were for a long time commonly accepted for the interpretation of experimental data on heat transfer in liquid flows.

Gases of similar atomic structure have a similar and constant value of the Prandtl number, so that, for such gases the heat transfer is described by

$$\mathrm{Nu} = c \cdot \mathrm{Re}^{m} \tag{1.2}$$

One of the first authors to apply the theory of similarity in heat transfer studies with cylinders in cross flow was Hilpert [12]. He also gave an interpretation of the heat transfer data in terms of Eq. (1.2) in his study with wires and tubes of different diameters in cross flow of air in the range of Re from 1 to $4 \cdot 10^5$. Fitting Eq. (1.2) to the experimental results suggested a variation of the value m, the power index of Re, from 0.33 to 0.805 for the range of Re covered.

Design calculations for boilers began to demand more exact data on the average heat transfer to cylinders in cross flows of gases. Because of this, further experiments were performed in flows of air, since its physical properties are so close to those of fuel combustion products.

Further insight in the process of the heat transfer was gained due to the studies of the effects of gas physical properties on heat transfer between a single cylinder and a gas crossflow. Mikheev [13] studied heat transfer in gases with different heat flux directions. From an analysis of his own results and the results of other authors, he concluded that the heat transfer coefficient was higher for wall-to-fluid heat transfer than it was for heat transfer in the opposite direction. His work on the interpretation techniques revealed a peculiar feature of the results accumulated at moderate thermal loads. By choosing the bulk flow temperature as a reference for the physical properties and for other determining parameters, the results of different sources could be approximated by a single curve.

Because of the need for a deeper insight into the heat transfer between fluid flows and solid bodies, further work was carried out, not only on the determination of the average heat transfer coefficient but also on measurement of local coefficients around the periphery of the bodies involved. The range of Re was extended for the first time by Kruzhilin and Shvab [14] in their studies on the local heat transfer from single cylinders in crossflow. Local heat transfer coefficients were also measured around a cylinder in air at high Reynolds numbers [15]. The results were later confirmed by Schmidt and Wenner [16].

In the front part of a cylinder, where there exists a laminar boundary layer, the heat transfer coefficient may be determined analytically. One of the earliest approximate solutions of this kind was obtained by Kruzhilin [17]. Later studies produced a number of calculation techniques, which involved analytical solutions of the boundary layer equations or of integral equations with the corresponding limiting conditions [18, 19]. As a result, sufficiently accurate heat transfer approximations are now possible for the front part of a cylinder, up to the separation point.

Recent advances in computer techniques are making feasible numerical heat transfer solutions for the whole circumference of a cylinder. So far, satisfactory results have been achieved only for lower values of Re [20].

All the above publications refer to the heat transfer and fluid dynamics in flows of air, and mainly to the subcritical range of the Reynolds number. Recent developments in science and technology–in particular in the chemical industries and in power generation (nuclear and conventional)–demand reliable equations of heat transfer between bodies and flows of viscous fluids. Research programs were initiated [21, 22] for still wider ranges of the Reynolds number. At the same time, it was important to tackle alternative aspects of the problem: the effects of fluid physical properties, of heat flux direction, and of the temperature variations had to be accounted for [23]. Under the initiative of one of the present authors (A.Z.), a research program was launched at the Institute of Physical and Technical Problems of Energetics, Academy of Sciences of the Lithuanian SSR. It was aimed at carrying out systematic studies of the heat transfer taking particular account of variable physical properties. The results of this program constitute the main part of this book.

The effects of fluid physical properties have been expressed in terms of equations containing the Prandtl number to an appropriate power index. Extensive studies in flows of air, water, and transformer oil suggest the value of 0.37 as the power index for Pr in the average heat transfer equations for the subcritical flow regime. However, for local heat transfer, the index varied from 0.33 to 0.45, with the minimum value in the front stagnation point, and the maximum in the rear vortex flow region.

The effects on heat transfer coefficient of the temperature difference and of the heat flux direction are caused by the variation of the fluid physical properties in the boundary layer. They may be accounted for by multiplying the coefficient calculated for constant physical properties by the ratio of the Prandtl numbers of fluid at the bulk and wall temperatures (Pr_f/Pr_w) to the power 0.25 temperature, provided the physical properties in the determining factors were referred to the free-stream temperature [22, 23].* Further studies revealed a lower intensity for wall-to-fluid heat transfer, to be reflected by the exponent 0.20.

For flow over a cylinder, there is a "critical" flow regime, which is obtained at high values of Re; this differs from the subcritical flow regime, in which a

*Editor's note: in Western literature the subscript f is sometimes used to denote "film temperature," i.e., a temperature midway between the bulk and wall temperatures. Here, it denotes the bulk or "free-stream" conditions.

laminar boundary layer exists on the front part of the cylinder and separates from the surface at $\varphi = 82°$. With the onset of the critical flow regime, a partial separation is observed on the front part. It is accompanied by a separation bubble. A turbulent boundary layer is then formed, which separates from the surface at $\varphi = 140°$. The local heat transfer coefficient distribution is closely related to this fluid dynamic structure. Thus, two minima are observed in the local heat transfer coefficient, corresponding to the partial separation and to the turbulent boundary layer separation, respectively.

Heat transfer studies on cylinders in crossflow of water and air [24, 25] revealed a further transition to the "supercritical" flow regime with a further increase of Re. Here the first heat transfer minimum is shifted upstream, while the second one remains practically stationary. This fluid dynamic transition is accompanied by a considerable augmentation of the heat transfer, and the results have been confirmed by Achenbach [26] in his studies with air.

The structure of fluid dynamics on bodies in crossflow featuring the separation bubble in the critical flow regime is discussed in the references [27, 28]. The early studies did not suggest any relation between the local variations of the heat transfer and the dynamics of the separation bubble. However, the interaction between the heat transfer and fluid dynamic processes was determined more exactly in the later special experiments.

In current studies of heat transfer from cylinders in crossflow, an increasing need is felt to take account of such factors as free-stream turbulence, blockage factor in the shell, surface roughness of the tubes, and the spatial flow structure in the wake. Perhaps the most important effect is that of free-stream turbulence, because of its relatively high levels. It ranges from 2-5 to 60% in some sorts of modern heat transfer equipment, and even in experimental rigs may be as high as 1%. The effects of turbulence may be highly important for accurate calculation of the heat transfer, mass transfer, and pressure drop in modern heat exchangers.

Among the earliest studies of the average heat transfer from cylinders, which included an account of the effect of turbulence, were those by Kirpichev and Eigenson [29, 30]. Later publications [31-35] contributed to the development of the theory of turbulent heat transfer. They referred to subcritical flows of air in wide ranges of Re.

An increase of free-stream turbulence has the effect of augmenting both the average heat transfer and its local values, although the effect is far from uniform around the cylinder circumference. The effect is most pronounced in the front zone. Higher turbulence leads to an earlier onset of the critical flow regime and to changes in the distribution both of velocity in the boundary layer, and of pressure on the cylinder surface. Certain peculiarities of the effect of turbulence were noted in viscous fluids. In general, systematic studies in turbulent liquid flows suggest that the heat transfer is highly complicated in character [36-38].

Turbulence measurements in most of the above studies were limited to those of longitudinal velocity fluctuations in the flow. A complete isotropy of the flow was assumed, and neither power spectra, nor micro- or macroscale turbulent structures were determined. This resulted in large discrepancies between the early results, such that consistent analysis or meaningful comparisons were hardly possible.

At present, the interpretation of the effects of a macroscale turbulence on the heat transfer is far from unanimous. Some authors do not find any significant effects of the turbulence, while others are inclined to claim a considerable influence, so that the problem is still open to further study. It also presents a challenge from the experimental point of view, because it calls for a simultaneous determination of the effects of both intensity and scale of the turbulence.

Both the heat transfer and the fluid dynamics are highly dependent on blockage factor (fraction of channel cross section occupied by the cylinders) in the channel in which the cylinders are fitted. Most practical implementations involve cylinders in flows that are also limited by the channel walls. The known experimental results refer to different blockage factors, with the exception of experiments performed in open sections of wind tunnels. Separate studies of the effects of the blockage factor are therefore necessary, since, through its effect on the velocity distribution outside the boundary layer, the blockage factor exerts an influence on the pressure distribution on the cylinder surface and on the velocity profile in the boundary layer. With high blockage factors, vortex shedding is no longer regular, and considerable changes occur in the parameters of the near wake, both pressure and friction coefficients being increased.

A common method of accounting for the effect of the blockage factor on the heat transfer is to modify the Reynolds number of referring it to different velocity values. Sometimes auxiliary coefficients are determined analytically from the velocity distribution in the channel, and introduced in the heat transfer equations. Those experimental and analytical studies of the blockage factor known to the present authors [39–41] all cover small changes of the blockage factor in subcritical flows of air with one possible exception [39]. Further studies in this area would be most welcome with more attention being paid to the combined effect, on the heat transfer, of turbulence and blockage factor. This combined effect has not (to the authors' knowledge) been yet considered.

Applications of augmented heat transfer from surfaces in cross flow have led to studies of rough-surface cylinders. Surface roughness destroys the laminar boundary layer and enhances the exchange of momentum and energy between the near-wall layer and the main flow. Experiments with rough-surface cylinders in crossflow of air [42] revealed a considerable augmentation of the heat transfer, particularly in the critical flow regime. However, the effect of fluid physical properties and of the Prandtl number on the heat transfer behavior of rough-surface cylinders is far from clear.

Both from the point of view of fluid dynamics and of the heat transfer, the rear part of a cylinder in crossflow presents a most complicated situation. For the complex vortex flow in this region, experimental studies and analytical solutions are both very difficult. Of the known approximations, perhaps that of Leontyev and Ryagin [43] is worth attention. It was based on the concept of boundary layer formation by steady-state vortices in the near wake. In the lower range of Re, the approach yields good qualitative results, but great discrepancies exist between the experiments and the analysis. A model of "vortex penetration" has also been suggested [44]. It assumes accumulation and regular transport of heat by vortices, which are periodically shed from the rear part of the cylinder.

Several studies suggest a close connection between the heat transfer and the processes of vortex formation and vortex shedding. Experiments in this area employed rotating cylinders [45] and auxiliary cylinders and splitter plates [46, 47] to eliminate regular vortex shedding. The manner of vortex shedding had a significant effect on the heat transfer of a cylinder, especially in the rear zone. Further studies of the rear part of a cylinder in crossflow are needed, because the available results are far from sufficient to form the basis of any analytical solution.

In the present state of knowledge, most emphasis is given to experimental studies. They are not only the source of still-scarce physical data, but also form the basis for compilation and development of analytical solutions. The priority is for structural determinations of fluid dynamic parameters both on the cylinder and in the wake.

Also welcome are attempts to systematize and to compare the accumulated information, for example reviews [7, 48] and compilations [49-51] of new calculation techniques for cylinders and other bodies in crossflow. In reference [49], data on the average heat transfer of different cylinders are approximated by a single curve, which is convenient for application in computer-aided design of heat exchangers.

The interpretation of the heat transfer and fluid dynamics on cylinders is discussed by Isachenko et al. [52] and the fundamental concepts of fluid dynamics and data on the heat transfer are reviewed by Schlichting [53]. But for a more profound understanding of the heat transfer from cylinders in crossflow, further experimental and analytical studies are necessary.

About a decade ago, a wide-ranging program of studies was launched at the Institute of Physical and Technical Problems of Energetics of the Academy of Sciences of the Lithuanian SSR. Similar studies were performed at the Institute of Technical Thermophysics of the Academy of Sciences of the Ukrainian SSR, and at the Institut für Reaktorbauelemente, KFA, Jülich, Germany.

The Lithuanian SSR program described in their book extended for a number of years and was aimed at exact determinations of the effects of fluid physical properties and fluid type on heat transfer. Local and average values of heat transfer were determined on circular, elliptic [111], and prismatic [137] tubes with different heat flux directions and different heat loads. The study covered the subcritical and the critical ranges of Re, and also the supercritical flow regime, which had never been studied before. Modern heat exchangers operate at maximum loads, under high effects of such factors as turbulence of different scales and levels, high blockage factors, and rough surfaces. All these factors have been duly considered in this study. The extensive experimental material presented refers to fluid dynamics around single cylinders, particularly in the near wake, to velocity distribution in the boundary layer, and to the distribution of pressure and friction. Heat transfer from a cylinder, especially from its rear part, was studied and related in detail to the fluid dynamic processes in the near wake. The same experimental rigs were used for tests in air, water, transformer oil, and aviation oil, so that the complete program covered the range of Reynolds number from 1 to $2 \cdot 10^6$, and of Prandtl number from 0.7 to 1000. The level of free-stream turbulence varied from 0.9 to 15% in each of the flows studied. The diameters of the circular, elliptic, and prismatic tubes varied from 0.4 to 90 mm.

In parallel with the experimental work, analytical solutions were performed for the local and average heat transfer using appropriate boundary layer equations taking into account the fluid physical properties.

We hope these numerous experimental and analytical results will provide a deeper insight into the fluid dynamics and heat transfer of cylinders. We believe that better understanding of the processes will become possible, and that the general interpretation techniques and other practical suggestions will be useful in the design of different heat exchangers.

CHAPTER

TWO

EXPERIMENTAL TECHNIQUES

The studies described in this book were carried out using specially designed rigs and experimental methods. To cover the effect of fluid type, the experiments were performed in flows of water, transformer oil, and air. For each of the fluids, fluid temperature was varied from 18 to 60°C, and wide ranges of the Reynolds number were covered. The range of Pr_f covered in the experiments was from 0.7 to 1000. By varying temperatures, velocity, and air pressure (up to 2.5 MPa), it was possible to vary Re_f from $4 \cdot 10^3$ to 10^6 in air, from $7 \cdot 10^3$ to $2 \cdot 10^6$ in water, and from $4 \cdot 10^2$ to $2.4 \cdot 10^4$ in transformer oil.

The following parameters were determined: local and average heat transfer, pressure coefficient distribution, friction drag, and pressure drag. The effects of free-stream turbulence, blockage factor, surface roughness, and fluid physical properties on the heat transfer and the dynamics of the boundary layer were determined.

To study the effect of the level of turbulence, artificial turbulence was introduced either into the free stream or into the near-wall space. Free-stream turbulence was varied from 0.2 to 15%.

The tests in the subcritical flow regime supplied information for comparisons with the results of other authors, and for the validation of our test rigs and methods.

For the tests with compressed air, modifications had to be introduced in the instruments and methods of measurement.

In interpreting the data, physical properties of water and air were taken from reference 136, and those of transformer oil were determined experimentally. Numerical values of the thermal physical parameters for the fluids studied are presented in Tables 1-5 in the Appendix. Minor differences in the properties of transformer oil were observed with different supply sources.

2.1 EXPERIMENTAL EQUIPMENT

For our experiments of fluid dynamics on cylinders in the subcritical flow regime ($Re < 2 \cdot 10^5$), five similar closed-loop rigs were constructed. The first was used for the circulation of transformer oil, the second for air at near atmospheric pressure, and the third for low flow rates of water. For the critical flow regime ($Re > 2 \cdot 10^5$), a wind tunnel capable of operating at air pressures up to 2.5 MPa was used. A new circulation loop was constructed for water. The rigs were practically identical, except for minor modifications. Figures 2.1, 2.2, and 2.3 show the arrangements of the loops for transformer oil, water, and air, respectively.

The test rigs for water and transformer oil each contained two pumps with different powers so that a wide range of flow rate (from 9 to 1000 kg/s) could be obtained to achieve the required range of Re.

In the high-pressure loop, air was circulated by a blast blower. In the low-pressure air loop, two axial fans of $8\text{-}10 \cdot 10^3$ m^3/h each were used. The high-pressure loop also incorporated a liquid separator.

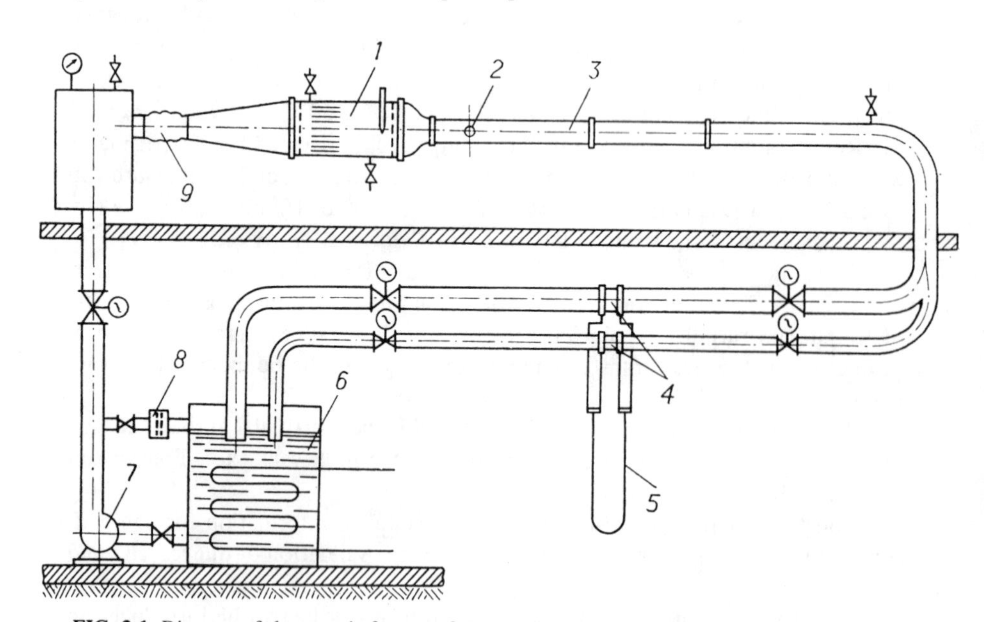

FIG. 2.1 Diagram of the test rig for transformer oil: 1, flow straightener; 2, test tube; 3, test section; 4, orifice plates; 5, differential manometer; 6, supply tank with a heat exchanger; 7, pump; 8, filter; 9, flexible junction.

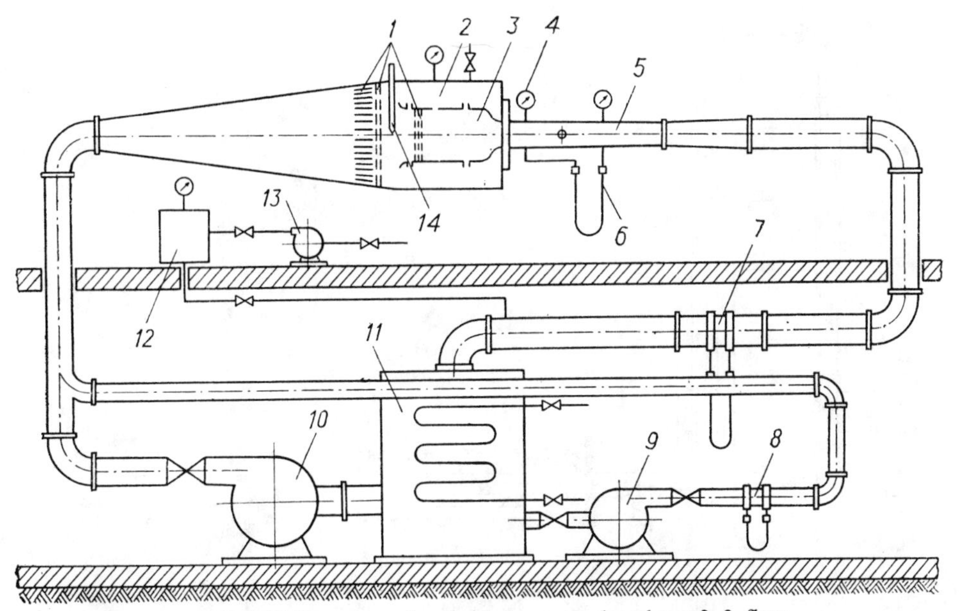

FIG. 2.2 Test rig for high-flow rates of water: 1, honeycomb and net; 2, 3, flow straighteners; 4, 6, manometers; 5, test section; 7, venturi tube; 8, orifice plate; 9, 10, pumps; 11, supply tank with a heat exchanger; 12, receiver; 13, anticavitation pump; 14, thermometer.

The flow rate was controlled either by values or by variation of the rotation velocity of the DC motor. The required Reynolds number was achieved by controlling the flow rate in the low-pressure air loop and by controlling pressure at constant-volume flow rate in the high-pressure loop.

The fluid was fed to a flow straightener with grids and screens, to ensure a uniform velocity profile. In the high-rate water loop, the flow straightener contained a honeycomb of aligned tubes 345 mm long and 15 mm in diameter in its entrance, to extinguish the large-scale turbulence. The small-scale turbulent structures were eliminated by two layers of a fine-mesh screen. The water was then passed to a smaller (300 mm × 300 mm) flow straightener containing two further layers of a fine-mesh screen. These screens, together with a diffuser preceding the test section, reduced the free-stream turbulence to a sufficiently low level. The other liquids loop contained similar sections for flow straightening, correction grids to maintain uniform velocity profiles, and additional honeycombs of aligned tubes 300 to 400 m long and 15 to 20 mm in diameter. Pitot tubes were mounted at the entrance of the test section to allow measurement of the velocity profile. Uniform velocity profiles were maintained throughout the experiments.

Total flow rate was determined using orifices and/or venturi tibes, measuring the pressure drop by either water or mercury manometer. Exact measure-

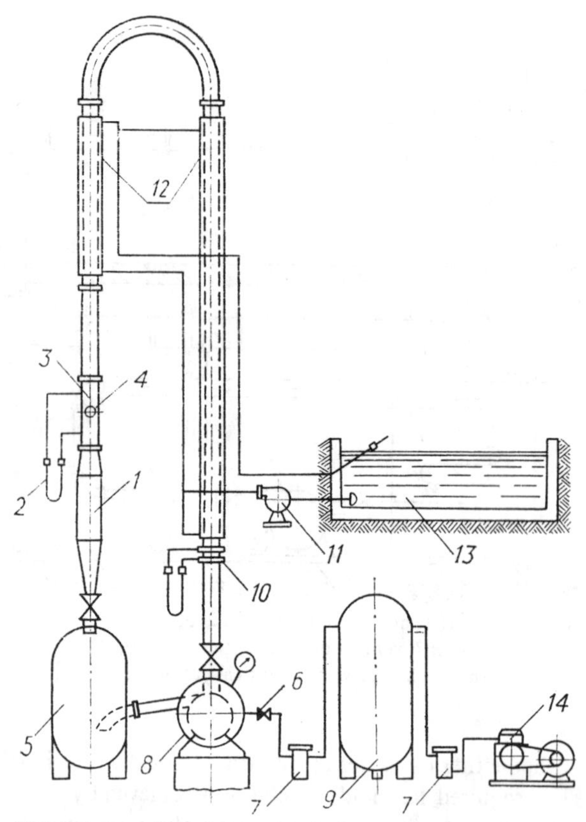

FIG. 2.3 Test rig for high-pressure air: 1, flow straightener; 2, differential manometer; 3, test section; 4, test tube; 5, collector–distributor vessel; 6, valve; 7, liquid trap; 8, blower drum; 9, receiver; 10, orifice plate; 11, pump; 12, cooling jacket; 13, storage tank; 14, compressor.

ment was ensured by the use of standard instruments. For low flow rates of transformer oil, double-orifice plates were used, and an auxilliary return pipe was introduced to check the measurement accuracy.

The air and liquids circulating in the loops were heated by the dissipation of kinetic energy. To maintain constant temperatures of the liquids, cooling coils were mounted in the storage tanks. The tanks also incorporated straightening baffles and deaeration screens. In the compressed-air loop, temperature was controlled by a cooling water jacket. In the low-pressure air loops, flow-straightening wedges, incorporated in the pipeline, were cooled to act simultaneously as cooling heat exchangers.

Bulk flow temperature was determined by thermocouples and thermometers in the entrance of the test section. The temperature was controlled to ±0.1°C for water and transformer oil, and to ±0.2°C for air. Appropriate insulation was provided for the storage tank and the pipelines to reduce heat dissipation from the loops.

In the large flow-rate water loop, anticavitation pumps were used in the test section and in the pipeline to provide a constant pressure of 1.6 MPa for different flow velocities. To exclude vibration, the pumps and the blowers were connected to the loop by flexible junctions.

A series of tests was also performed in an essentially infinite flow of water, in an open lake. The test cylinder was suspended on a special support from beneath a boat to a depth of 0.9 m.

The test sections Four separate test section channels of 100 × 200 mm, 150 × 180 mm, 150 × 101 mm, and 200 × 200 mm cross sections were used. Each of them was constructed as a rectangular stainless steel channel with an acrylic resin window for visual observation. They were constructed of integral interchangeable pieces, and the different cross section test sections were remounted quite easily.

Figure 2.4 shows, as an example, the test channel used for experiments with circular smooth-surface cylinders in water in the critical flow regime ($Re > 2 \cdot 10^5$). The cylinders were mounted 670 mm from the entrance, preceded by traversable Pitot probes. These probes were located horizontally and vertically in a single plane at 160 mm from the entrance. Static pressure was measured through two openings of 0.3 mm diameter in the opposite walls of the test channel connected by pressure hoses.

A special test section was used for the heat transfer studies in air at the critical range of Re. It was 200 × 200 mm wide and 1200 mm long, and had two 0.3-mm static pressure gauges inserted through the opposite channel walls to traverse the front critical point of the test cylinder.

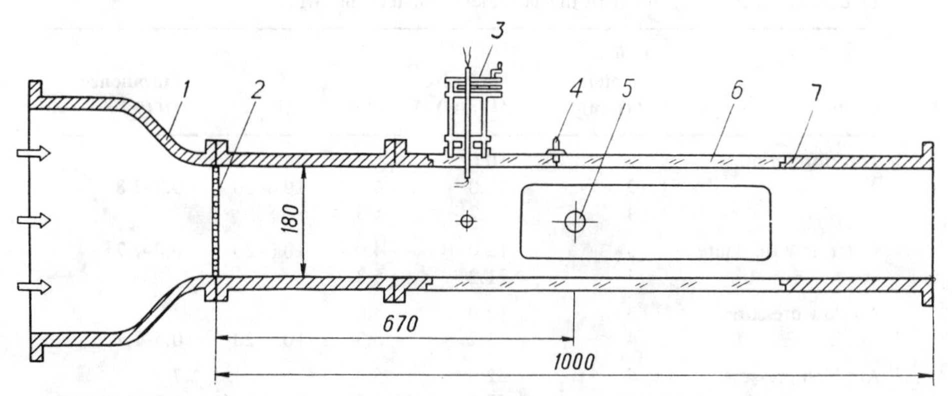

FIG. 2.4 Test section for fluid flow and heat transfer studies of a cylinder: 1, diffuser; 2, turbulence generator; 3, traversing device; 4, static pressure gauge; 5, test tube; 6, acrylic resin window; 7, test section.

The level of turbulence in the test sections was considerably reduced by the multilayer fine-mesh screens in the flow straighteners. Whenever high isotropic free-stream turbulence was needed, it was generated artificially by grids preceding the test section. Their parameters are shown in Table 2.1.

High levels of turbulence ($Tu = 15\%$) in the high-pressure wind tunnel were also generated artificially using a honeycomb of 20-mm-diameter circular rods at a distance of 160 mm from the front stagnation point of the test tube. The stainless-steel circular rods were fixed in a frame at different distances M from one another. The frame could be located in the test section at different distances from the test tube, to control free-stream turbulence. Using these frames it was possible to control the micro and macro scales (Λ and L) of the turbulence in the ranges 1.5 to 10 mm and 8 to 25 mm, respectively.

2.2 DETERMINATION OF HEAT TRANSFER FROM A CYLINDER

Local heat transfer was studied using electrically heated tubes. Measurements were made of heat fluxes and the surface temperature, and values of the local heat transfer coefficient were determined from

$$\alpha_x = \frac{q_w}{t_{wx} - t_f} \tag{2.1}$$

where q_w is heat flux from the tube surface and t_{wx} is the local wall temperature obtained from the thermocouple reading.

Having determined the distribution of the local heat transfer coefficient around the cylinder, the average heat transfer coefficient for the whole cylinder

TABLE 2.1 Major Parameters of the Turbulence Generating Grids

Loop	Grid diameter d (mm)	Mesh size M (mm)	M/d	x/M	Turbulence, Tu (%)
Transformer oil	2	10.6	5.3		
	3	12.0	4.0	10.6–20	0.7–7.8
	4	21.2	5.3		
Water, low flow rate	3	12.0	4.0	10.6–20	0.8–7.7
	4	21.2	5.3		
Air, low pressure	3	12.0	4.0		
	4	21.2	5.3	10.6–20	0.5–7.7
Air, high pressure	6	32.0	5.3	7	7
	20	40.0	2.0	4	1.2–15
Water, high flow rate	10	19.0	1.9	35	1.0–7

could be determined. The simplest way of doing this is from the mean integral value of the local coefficient

$$\alpha = \frac{1}{l} \int_0^l \alpha_x \, dx \tag{2.2}$$

However, with highly nonisothermal surfaces and high temperature differences, the average heat transfer coefficient was determined more accurately by dividing the integral mean value of the heat flux by the integral mean value of the surface temperature:

$$\alpha = \frac{\frac{1}{l} \int_0^l q_w \, dx}{\frac{1}{l} \int_0^l \Delta t_x \, dx} \tag{2.3}$$

The latter determination technique is more time-consuming, but it yields a better agreement with the average heat transfer obtained with isothermal surfaces (constant t_w). Both methods were used in the processing of the present data.

In addition to the electrically heated tests, average heat transfer coefficient was also determined using water calorimetry with a nearly isothermal surface. The average heat transfer coefficient was found from

$$\alpha = \frac{Q}{F(t_w - t_f)} \tag{2.4}$$

where the heat flow Q was measured calorimetrically (i.e., by measuring the difference between the inlet and outlet temperatures and flow rate of the water in the inside of the tube), and the wall temperature t_w was found as a mean integral value of the thermocouple readings on the cylinder surface. F is the outside surface area of the tube.

The major parameters of the tubes used in the water-calorimetry tests (i.e., diameters d, length l, and the height of the test sections H) are presented in Table 2.2.

By using tubes of different diameters, the effects of the blockage factor on the heat transfer could be evaluated.

The electrically heated tubes were of similar construction (Fig. 2.5), although the diameter was varied. The tube was formed around a textile-reinforced plastic rod, with grooves for the thermocouple connections, whose surface was covered with a 0.1-mm-thick constantan foil that was heated by current fed through the copper end pieces as shown. For high-velocity flows of water the constantan was replaced by a more resistant stainless-steel tube whose wall thickness was 0.25 ± 0.005 mm. The low thermal diffusivity of the steel

TABLE 2.2 Major Parameters of the Test Channel and of the Water-Calorimetry Test Tubes

d (mm)	l (mm)	l/d	H (mm)	d/H
12.0	100	8.3	50	0.24
30.0	100	3.3	200	0.15
30.1	150	5.0	101	0.30
30.7	150	4.9	180	0.17
32.0	200	6.2	200	0.16
50.0	150	3.0	180	0.27
50.0	200	4.0	200	0.25
50.0	150	3.0	101	0.50
50.0	100	2.0	200	0.25
50.2	100	2.0	200	0.25
68.4	150	2.2	101	0.68

tubes and of the constantan foils, combined with their constant thickness and the low thermal conductivity of the plastic rod, guaranteed a sufficiently constant heat flux on the surfaces studied. To measure the surface temperature, copper-constantan thermocouples were welded to the internal surfaces of the tubes. The thermocouples were 0.1 and 0.12 mm thick for the constantan foil cylinders, and 0.18 mm thick for the steel tubes. They were located at the central cross section of the cylinder at 10° intervals, and covered 190° arcs on the cylinder surfaces.

By rotating the test tube on its horizontal axis, measurements were performed at 5° intervals. For the critical flow regime in water, the surface temperature fluctuated so average values were taken from three readings on each of the thermocouples. In addition to the thermocouples at the central cross section, others were fixed at intervals along the cylinder, and used to estimate the axial heat loss and also to measure the voltage drop in the middle part of the cylinder,

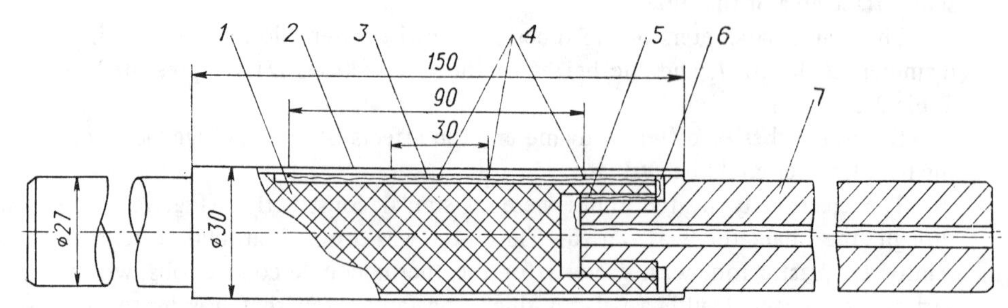

FIG. 2.5 Electrically heated cylinders used for the measurement of the local heat transfer: 1, textile-reinforced resin rod; 2, heat transfer surface of constantan or stainless steel; 3, thermocouple conductors; 4, thermocouples and contacts for the measurements of the voltage drop; 5, insulation collar; 6, thermocouple leads; 7, copper contact.

where all the measurements were performed. The heat transfer coefficients measured refer to this central zone, where end losses were insignificant. The thermocouple leads were led out through an end-collar and a seal, which prevented fluid leakage.

Surface temperature was calculated from the thermocouple readings, taking account of the temperature drop in the wall using the equation

$$\Delta t = \frac{q_v r_1^2}{4\lambda} \left[\left(\frac{r_2}{r_1} \right)^2 - 2 \ln \frac{r_2}{r_1} - 1 \right] \tag{2.5}$$

where q_v is the volumetric heat release rate (W/m^3) in the wall.

Two separate test tubes were constructed for the determination of the surface roughness effect. They were both 30 mm in diameter and had surfaces covered with a 0.2-mm-thick constantan foil knurled or punched to produce artificial roughness.

The first surface, simulating the random character of a sand-roughened surface, had pyramidal protrusions introduced by a special roll prepared by side knurling, the foil being fed to the roll together with a cardboard support; 0.15-mm-high pyramidal elements were formed on its surface. The pyramids were rhombic in form, with 60° and 120° angles and with 0.5-mm distances between parallel sides (see Fig. 2.6). The foil was fixed onto a textile-reinforced plastic rod, which had longitudinal grooves for the thermocouple leads.

The temperature of the rough surfaces was determined in a manner similar to that for the smooth surface. In high-velocity water flows, the surface temperature was subject to additional fluctuations because of vortex regions formed downstream of each of the surface roughness elements.

The second roughened surface was also covered with pyramids of form similar to those of the first surface, which were 1.2 mm high and located 3.56 mm from each other. In this case, the 0.2-mm-thick constantan foil was shaped by punching using a special pair of dies.

Surface temperature was measured at the apex of the pyramids. However, no significant difference was noted in comparing temperature measurements at different locations.

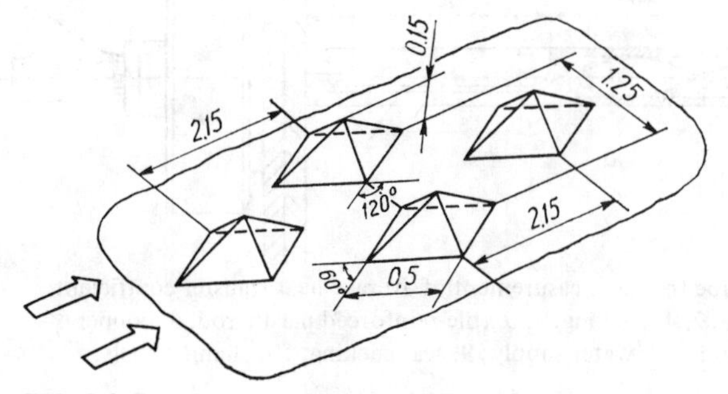

FIG. 2.6 Rough surface form used (dimensions refer to the first surface).

Both of the rough-surface cylinders satisfied the condition of a constant heat flux (q_w = constant). In calculating the heat flux, the surface area of the pyramids was included in the total area of the heat transfer surface.

Calorimeter tubes were also used in the determination of average heat transfer coefficient. The copper water calorimeter (Fig. 2.7) was 12 mm in diameter and 100 mm in length. Its ends were sealed with bushes of textile-reinforced plastic. Proper mixing of the water inside the test tube was facilitated by a helical coil. A steady-state supply of water was ensured using a thermostatically controlled tank and a set of compensation tanks. The rate of heat release was determined from

$$Q = Gc_p\,\Delta t \tag{2.6}$$

The temperature drop Δt inside the calorimeter tube was measured by a four-junction differential thermocouple. The water flow rate G was found by weighing. The average heat transfer coefficient was determined using the readings of thermocouples welded to the external surface of the calorimeter.

An electrically heated calorimeter tube was also used to measure the average heat transfer coefficient. The construction and the location of thermocouples on this tube were similar to the water calorimeter, but the heat was generated using a nichrome coil inside the tube. The nichrome coil was mica-insulated to maintain uniform heating, to minimize end heat losses, and to prevent shorting between adjacent coils. The voltage drop along the coil was measured in the central zone of the tube. The amount of heat emission was determined from the voltage drop ΔU and from the amperage I:

$$Q = \Delta UI \tag{2.7}$$

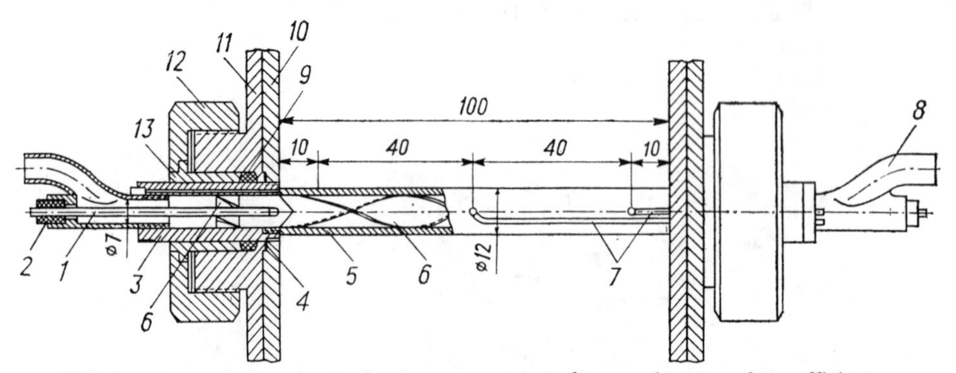

FIG. 2.7 Water calorimeter tube for the measurement of average heat transfer coefficient: 1, four-junction thermocouple; 2, 4, packing; 3, textile-reinforced plastic rod; 5, copper cylinder; 6, coil; 7, thermocouples; 8, water supply; 9, seal packing; 10, channel walls; 11, seal; 12, nut; 13, collar.

Current was supplied from a DC generator through an electronic stabilizer. The average tube surface temperature was determined from the mean of the thermocouple readings in the central zone.

In all experiments with water, bulk flow temperature was measured using copper-constantan thermocouples in the flow straightener, and checked on a thermometer with 0.1° divisions.

The voltage drops and thermocouple emf values were usually measured with a digital millivoltmeter linked to Solartron instrumentation. In some cases, the emf of the thermocouples was determined using an accurate portable potentiometer. Exact calibration of the thermocouples preceded the experiments.

To exclude voltage pick-up, the emf of the thermocouples in a direct contact with the calorimeter surface was determined as an arithmetical mean of two readings for the electric current flowing in one direction and the other, respectively.

Heat flux on the surface of the tube was found from

$$q_w = \frac{I\Delta U}{F} \tag{2.8}$$

where I is amperage, ΔU is resistance drop in the test part of the calorimeter, and F is total area of the heat transfer surface.

The test tubes were heated electrically from low-power DC generators through constant-resistance leads so that the heat flux could be controlled within 1%. A shunt and a number of millivoltmeters were included in the circuit to control the electrical power.

Local heat transfer coefficients could also be determined from the electrical calorimeter tube results, if account was taken of lateral heat conduction along the surface caused by the temperature gradient. Such determinations covered the whole perimeter of the tube. The lateral heat flux values were determined from the second derivative of the surface temperature by

$$q = \lambda \frac{\partial^2 t}{\partial x^2} \delta \tag{2.9}$$

where δ is the wall thickness and λ the thermal conductivity of the wall. The effective local heat transfer coefficient was given by

$$\alpha_x = \frac{q_w \pm q}{t_{wx} - t_f} \tag{2.10}$$

The second derivative of the surface temperature was computer-determined from the temperature distribution data by a special program. The lateral fluxes amounted to 30% and to 1-2% of the total for flows of air and water, respectively.

2.3 DETERMINATION OF FLUID DYNAMIC PARAMETERS

Pressure distribution on the cylinder surface was measured with the help of special test tubes. Four stainless-steel tubes of similar construction and 30, 50, and 105 mm in diameter were manufactured (Fig. 2.8). The variation of diameter allowed the determination of the effect of shell blockage factor on the pressure distribution. The cylinders were smoothly polished and had 0.3-mm-diameter openings at their central cross section. The openings were connected by high-pressure hoses to U-tube manometers of mercury or water, depending on the pressure. Static pressure was measured through two 0.3-mm-diameter holes, located in the opposite walls of the channel and connected by high-pressure hoses.

To determine circumferential distribution of pressure, the test tube was rotated on its horizontal axis, and readings were taken at intervals of 2 to 5°. The results were expressed in terms of velocity heads:

$$\bar{p}=\frac{p_{\varphi}-p_{\infty}}{\frac{1}{2}\rho U_{\infty}^{2}} \tag{2.11}$$

The fluid velocity outside the boundary layer was determined around the circumference by applying the Bernoulli equation

$$U_{\varphi}=U_{\infty}\sqrt{1+\frac{2\left(p_{\infty}-p_{\varphi}\right)}{\rho U_{\infty}^{2}}} \tag{2.12}$$

to the data for circumferential pressure distribution.

For the measurements of circumferential pressure distribution on rough-surface cylinders, two further test tubes were prepared by covering smooth-surface cylinders with a constantan foil bearing the pyramidal elements described in the preceding section. The readings of pressure were taken through two 0.3-mm-diameter holes located at the central cross section, one on the top of a

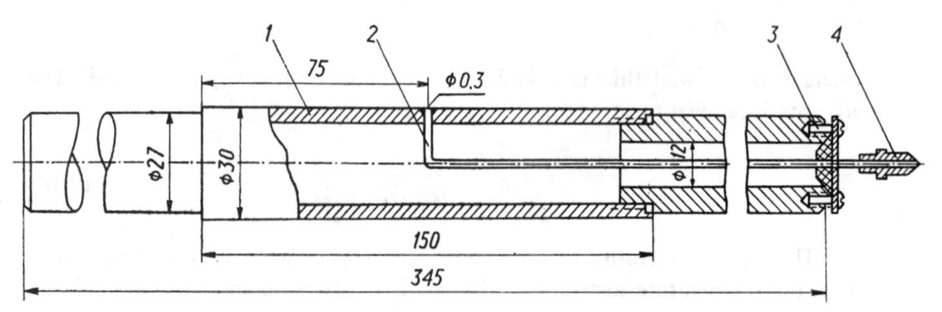

FIG. 2.8 Test tube for the measurement of the pressure distribution: 1, tube wall; 2, pressure gauge; 3, seal; 4, end piece.

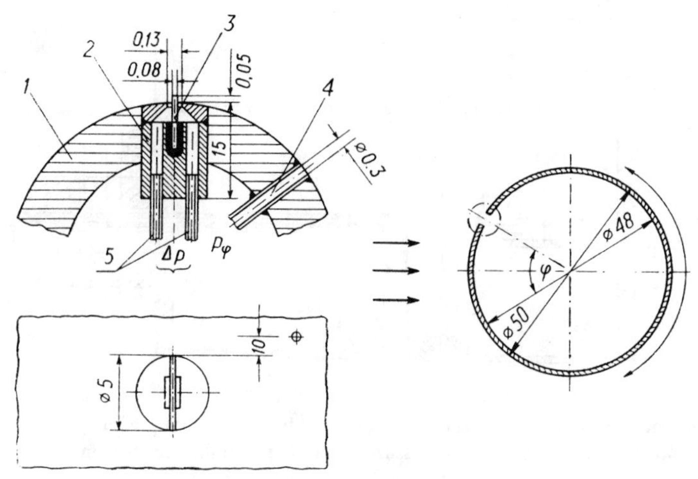

FIG. 2.9 The protruding-fence skin friction probe: 1, tube wall; 2, probe body; 3, fence; 4, surface pressure tapping; 5, pressure connections to upstream and downstream sides of fence.

surface element and the other on the base tube. No significant pressure differences were noted between the two tappings for the small-scale roughness, although this was not the case with the surface having the larger elements.

Wall shear stress (skin friction) was measured on a 50-mm-diameter smooth cylinder using a surface-fence skin friction probe of the type designed by Konstantinov and Dragnysh [54] (Fig. 2.9). The technique is based on the direct relation between the shear stress τ_w and the pressure difference Δp between the upstream and downstream sides of the fence:

$$\tau_w = a_1 \, \Delta p \tag{2.13}$$

The arrangement of the tube used for the shear stress tests is shown in Fig. 2.10; the skin friction probe was glued to the internal surface as shown. The probe was flush with the surface, any unevenness smoothed out with epoxy resin, with just the fence protruding. The 0.08-mm-thick fence of stainless steel was fixed on a brass plate and a special calibration device was used to adjust its protrusion in the range 0.015 to 0.005 mm, thus allowing the probe to cover a wide range of Re.

The test tube also contained two 0.3-mm-diameter pressure tappings (only one shown in Fig. 2.10) on the same side as the fence probe and, on the opposite side, two further slot tappings (1 × 0.03 mm cut along the cylinder axis).

Calibration of the skin friction probes was performed in the following manner. Using the Bernoulli equation [Eq. (2.12)] the velocity outside the boundary layer U_φ was calculated from the measured pressure distribution. The functions $U_\varphi/U_\infty = f(\varphi)$ was determined around the whole circumference. These functions were used to evaluate the shear stress distribution in the laminar

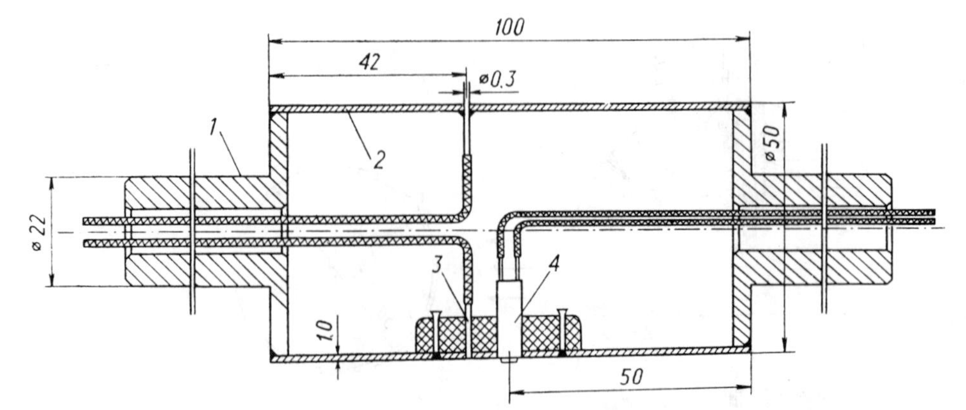

FIG. 2.10 Test cylinder used for the measurement of the skin friction and pressure distribution: 1, support; 2, cylinder wall; 3, pressure gauge; 4, protruding-fence skin friction probe.

part of the boundary layer (up to $\varphi \approx 40°$) using a previous calculation technique [55], which was based on a momentum equation. Thus, it was possible to construct curves of the form $\tau_w = f(\varphi)$ and using the direct proportionality between τ_w and Δp, the pressure drop measured across the fence, direct calculation functions $\tau_w = f(\Delta p)$ were found (Fig. 2.11). Similar techniques were used to find calibration functions for flows of air, water, and transformer oil and for the protrusion levels of the surface fence causing the least possible degree of turbulence while still allowing accurate evaluation of τ_w.

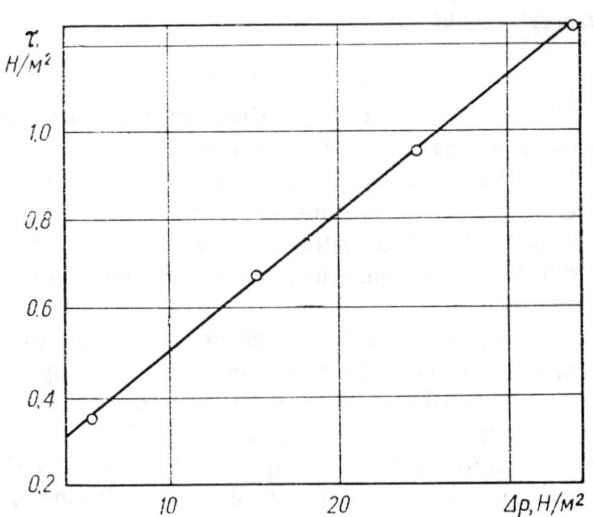

FIG. 2.11 Calibration function for the protruding-fence measurement of the shear stress (water, $Re = 5.12 \cdot 10^4$).

The pressure distribution on the tube surface and pressure drop on the surface fence were measured by micromanometers or by U-tube manometers, depending on the magnitude of the pressure drop.

Free-stream turbulence was measured by constant-temperature thermal anemometers whose response was analyzed using DISA instrumentation.

The sensor of the thermal anemometer is maintained constant by balancing a Wheatstone bridge using a servoamplifier (Fig. 2.12). The current from the servoamplifier is fed through the bridge and to the sensor contacting the fluid. The sensor is made of metal with a high temperature coefficient of electric resistivity and heated through a variable resistance R_3 up to a controlled temperature t_w. For constant pressure, temperature, and fluid physical properties, the amount of heat transferred from the sensor is governed only by fluid velocity. Any change of fluid velocity would tend to change the temperature, and thus the electrical resistance, of the sensor, and the voltage between a and b is changed accordingly to maintain t_w. The polar properties of the servoamplifier are designed to give an increase of voltage (and current) fed to the Wheatstone bridge whenever the sensor temperature t_w is lowered and its resistance increased. The original temperature is restored immediately, so that the temperature of the sensor remains constant, and velocity fluctuations in the fluid are reflected in the corresponding voltage variations in the amplifier at c.

For measurements in air, a hot-wire sensor was used. The wire was of platinum-tungsten, 5 μm in diameter and 1.2 mm in length ($l/d = 240$). Measurements in transformer oil were performed with sensors made of nickel foil, fixed on V-shaped glass supports, and insulated by a 2-μm layer of quartz.

Sensors for water flows were similarly of nickel foil fixed on glass cones and insulated by a layer of quartz.

The overheat ratios for the sensor were 0.7, 0.1, and 0.3 for air, water, and transformer oil, respectively.

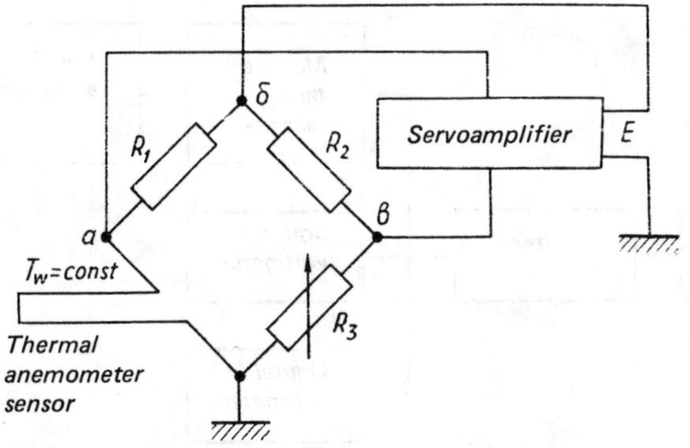

FIG. 2.12 Measurement circuit of a constant-temperature anemometer: 1, sensor; 2, servoamplifier; $\delta - b$, $b - c$.

The relationship between sensor heat loss Q and velocity U_∞ was in the form

$$Q = A + B(U_\infty)^n \tag{2.14}$$

where A and B are constants for a given fluid, temperature, and pressure, and the exponent n varies from 0.4 to 0.5. Calibration of the sensors was performed on a calibration facility (for air), and in the test sections themselves for water and transformer oil. A linearizer was included in the measuring circuit to produce a linear dependence of the output signal on the flow velocity, thus facilitating turbulence measurement (Fig. 2.13). Free-stream turbulence was measured in the test section with the test cylinder removed, in the plane normal to the flow and passing through the front stagnation point. In the whole range of Re studied, a nearly constant level of turbulence was maintained. Some difficulties were encountered for water flow in the critical region, where the turbulence decreased with increase of velocity. However, when the turbulence level was high, constant turbulence could be assumed throughout the flow velocity range (Fig. 2.14).

Before commencing an experiment, a check was made of the uniformity of turbulence in the test section, both with and without the honeycomb. The sensor of the thermal anemometer was traversed across the channel using the traversing device. Sufficiently uniform distribution of turbulence was observed.

To evaluate the scale of turbulence, magnetic tape records were made (over a period of 3-4 min) of the anemometer signal (Fig. 2.13) and subjected later to a statistical analysis on a Fourier analyzer. To achieve stable correlation functions and spectra, samples of 25600 successive instantaneous values were collected in each of the flow regimes. The large samples had smooth correlation functions and supplied sufficient data to obtain information about the long-time-scale events in the flow. The short-time-scale events could be observed by averaging data for shorter sampling periods. The statistical analysis gave data for spectra, time correlation functions, and for micro- and macroturbulent structures.

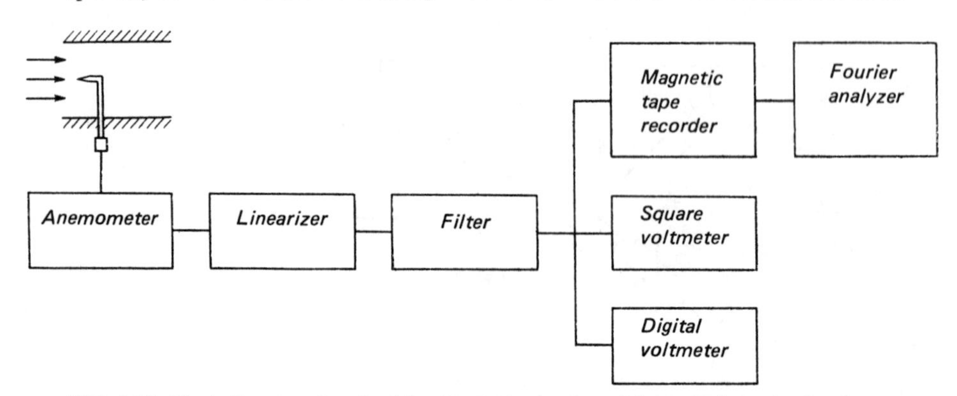

FIG. 2.13 Block diagram of method for the determination of the turbulence level and the turbulence scale.

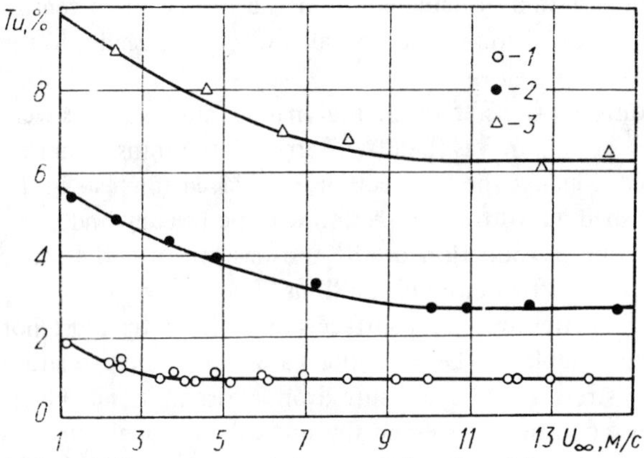

FIG. 2.14 Free-stream turbulence as a function of free-stream velocity in water flow at high velocities: 1, with a fine-mesh screen; 2, natural turbulence in the loop. 3, with a turbulizer grid of 9-mm-diameter, 17-mm-long aligned rods.

The Euler coefficient of the time correlation was determined from

$$R_{x(\tau)} = \frac{\overline{u'(\tau)\, u'(\tau+\Delta\tau)}}{\bar{u}'^2} \tag{2.15}$$

and averaged for time τ. The Euler integral time scale was then found,

$$I_x = \int_0^\infty R_{x(\tau)}\, d\tau \tag{2.16}$$

and the spatial integral scale,

$$L = U_\infty I_x \tag{2.17}$$

The Euler microscale of time τ_x was given by

$$\frac{1}{\tau_x^2} = -\frac{1}{2}\left[\frac{\partial R_x(\tau)}{\partial \tau^2}\right]_{\tau=0} \tag{2.18}$$

and the spatial microscale

$$\Lambda = U_\infty \tau_x \tag{2.19}$$

An analysis of the results for water suggested a growth of both micro- and macroscale turbulence with free-stream velocity. The tendency was also observed in other fluids.

2.4 NOTES ON EXPERIMENTAL PROCEDURE

Each experiment was preceded by a careful Pitot tube check of the velocity distribution in the channel, to ensure uniformity. The bulk flow velocity was then compared to the velocity value calculated from the orifice plate venturi tube readings. After this check was completed, the tube was introduced into the test section.

For a given test section, heat transfer and fluid dynamic measurements were first performed in air; the test section was then transferred to the transformer oil loop and the tests repeated. Finally, the test section was placed for tests in the water loop. The oil was washed out of the test section with petroleum, and it was properly dried out before the introduction of the test section into the water loop. This excluded the possible effect of emulsion films.

In measurements of skin friction with a surface fence, the latter must not protrude beyond the viscous sublayer. Only in this case can a linear relation between the surface shear stress and the pressure drop across the fence be assumed. Conforming to this condition was especially difficult on rough surfaces at high Re.

A further complication was introduced by the effect of electrolysis. On test tubes heated electrically from the DC generator, the surface thermocouples failed after a short usage, especially at high Re. Their viability had to be checked out before each series of tests. For this purpose, thermocouple readings in different locations on an unheated tube were compared to each other, and to those in the flow straightener. To reduce the errors in the data, at least three values of the digital millivoltmeter output were taken and averaged for each reading.

The symmetry of the flow around the test tube was continued by measurement of temperature or pressure around the circumference for a complete cycle of rotation. The flows were found to be symmetrical throughout.

An error analysis that included the effects of the measurement techniques, the instruments used, and the sampling intervals showed the error bands for the thermal and fluid dynamic parameters were in the range of 1.7 to 4.5% for water flows, 2 to 5.0% for air flows, and 1.9 to 4.8% for transformer oil flows.

CHAPTER

THREE

ANALYTICAL TECHNIQUES

In the range of Re from 10^5 to 10^7, the laminar, transitional, and turbulent boundary layers are observed simultaneously on a cylinder. This highly complicated fluid dynamic pattern has been a stumbling block in the development of analytical descriptions of the heat transfer in this range. However, there are a limited number of analytical studies for the subcritical flow regime.

Any attempt at formulating the problem involves as its first step the choice of the initial equations. Numerous authors choose the Navier–Stokes equations for a complete description of the velocity field; combined with the energy equations, they give a most general solution of the convective heat transfer. However, general nonlinear differential equations of this kind have no analytic solutions. Various finite-difference solutions have been obtained for Re up to 500, such as those by Eckert et al. [20, 56], and Lin et al. [57]. A more thorough analysis of the choice of initial equations was performed by Cebeci and Smith [67]; they preferred to extend the boundary-layer equations by adding heat transfer equations and by making a number of hypotheses and assumptions. Even simpler, indirect, solutions of the boundary-layer equations have also been used for the heat transfer; for example, Eckert [19] gives a polynomial expression for the velocity and temperature profiles at Re up to $1.7 \cdot 10^5$.

Traci and Wilcox [58] and Smith [59] attempted to evaluate the effect of free-stream turbulence on the heat transfer in the region of the front stagnation point, but their results apply only to Pr = 0.7.

The effect of blockage factor k_q on heat transfer from a grid of cylinders was studied by Kochin [60] and Borisenko [61], and series descriptions were used for the velocity distribution in the outer part of the boundary layer. A model accounting for k_q was suggested in reference 39 for a single cylinder, in which the velocity distribution for the front part of the cylinder approached the unblocked form for $k_q = 0.3$.

In this chapter we present a solution for the boundary-layer equations in the laminar, transitional, and turbulent parts of the boundary layer, taking into account effects of Tu, k_q, and Re on the heat transfer and fluid dynamics for a cylinder in crossflow.

3.1 THE INITIAL SYSTEM OF EQUATIONS

If the pressure gradient across the boundary layer is neglected, the following system of equations can be used for the fluid dynamics and heat transfer in a two-dimensional boundary layer of a viscous incompressible fluid:

$$\rho\left(u\frac{\partial u}{\partial x}+v\frac{\partial u}{\partial y}\right)=\frac{\partial \tau}{\partial y}-\frac{\partial p}{\partial x} \tag{3.1}$$

$$\frac{\partial(\rho u)}{\partial x}+\frac{\partial(\rho v)}{\partial y}=0, \tag{3.2}$$

$$c_p\rho\left(u\frac{\partial T}{\partial x}-v\frac{\partial T}{\partial y}\right)=-\frac{\partial q}{\partial y} \tag{3.3}$$

with the boundary conditions

$$y=0,\ u=v=0,\ T=T_w \quad \text{or} \quad q=q_w \tag{3.4}$$

$$y=\infty,\ u=U,\ T=T_f \tag{3.5}$$

3.2 THE LAMINAR BOUNDARY LAYER ON THE FRONT PART OF A CYLINDER

The analysis was performed for $10^4 \leqslant \mathrm{Re} \leqslant 10^5$. The system of Eqs. (3.1)-(3.3) was closed using an empirical expression for the velocity distribution in the outer boundary layer, due to Hiemenz [82] and reported by Eckert [19], which has the form:

$$\frac{U}{U_\infty}=3.6314\frac{x}{d}-2.1709\left(\frac{x}{d}\right)^3-1.5144\left(\frac{x}{d}\right)^5 \tag{3.6}$$

The velocity was related to the longitudinal pressure gradient by the Bernoulli equation,

$$-\frac{dp}{dx}=\rho U\frac{dU}{dx} \tag{3.7}$$

The theory of differential equations indicates that a partial derivative solution for the parabolic Eqs. (3.1)-(3.3) [with the given boundary conditions of Eqs. (3.4) and (3.5) and the closing condition of Eq. (3.6)] is only possible for the region located downstream of an initial cross section x_0 for

$$x = x_0, \quad u = u_0(y), \quad T = T_0(y) \tag{3.8}$$

The initial velocity profiles were chosen by the method of Holstein and Bohlen [72]:

$$\frac{u}{U} = (2\eta - 2\eta^3 + \eta^4) + \frac{\Lambda}{6}(\eta - 3\eta^2 + 3\eta^3 - \eta^4) = F(\eta) \tag{3.9}$$

where $\eta = y/\delta$ (δ being the hydrodynamic boundary-layer thickness) and where $\Lambda \approx 7$ at $\varphi = 1°$. The temperature profiles were described by a polynomial suggested by Squire [73]:

$$\frac{T - T_f}{T_w - T_f} = 1 - F(\eta_T) \tag{3.10}$$

where $\eta_T = y/\delta_T$, where δ_T is the thermal boundary-layer thickness. The values of δ and δ_T were determined as in [72] and [73]. Taking $\Lambda = 7$, the boundary-layer thickness is

$$\delta = \sqrt{42.88 \frac{\nu}{U^6} \int_0^x U^5 dx} \tag{3.11}$$

The ratio $\delta_T/\delta = \Delta$ can be determined from

$$\Delta^2 H(\Delta) = 0.093 \frac{U^4}{\mathrm{Pr}} \frac{\int_0^x UH dx}{H \int_0^x U^5 dx} \tag{3.12}$$

and for constant Δ from

$$\Delta^2 H(\Delta) = 0.093 \frac{U^4}{\mathrm{Pr}} \frac{\int_0^x U dx}{\int_0^x U^5 dx} \tag{3.13}$$

$H(\Delta)$ is given by:

$$H(\Delta) = \frac{2}{15}\Delta - \frac{3}{140}\Delta^3 + \frac{1}{180}\Delta^4 \quad \text{when} \quad \Delta < 1 \tag{3.14}$$

or

$$H(\Delta) = \frac{3}{10} - \frac{3}{10 \cdot \Delta} + \frac{2}{15 \cdot \Delta^2} - \frac{3}{140 \cdot \Delta^4} + \frac{1}{180 \cdot \Delta^5} \quad \text{when} \quad \Delta > 1 \tag{3.14a}$$

Calculating the value of $H(\Delta)$ from Eq. (3.14) or (3.14*a*), the value of Δ can then be found from Eq. (3.12). Successive iteration is employed, using $\Delta =$ constant $= 1$ as the first approximation. The value of $H(\Delta)$ from Eq. (3.13) is then substituted in the right-hand side of Eq. (3.12) and gives the next approximation of Δ. Usually five iterations give sufficiently accurate approximation. The value of δ_T is then determined from the calculated Δ and using the value of δ calculated from Eq. (3.11). An interesting observation is that for φ in the range 0 to 10°, $\Delta = 1.57$ for Pr = 0.7, and $\Delta = 1$ only for Pr = 2.57. A similar observation was made by Kruzhilin [17].

For solution for q_w = constant, an important initial condition is the value of T_w at cross section x_0 (the front stagnation point); this must be determined separately for each case. It can be found with sufficient accuracy from the equation suggested by Eckert [19] for the front stagnation point,

$$\mathrm{Nu}_{fd} = 1.14\, \mathrm{Re}_{fd}^{0.5}\, \mathrm{Pr}_f^{0.35}\, (\mathrm{Pr}_f/\mathrm{Pr}_w)^{0.25} \tag{3.15}$$

To facilitate comparison of experimental and analytical results, the value of T_w can be determined from the measured local heat transfer values,

$$\alpha_x = \frac{q_w}{T_w - T_f}$$

The solution starts on the front part of the cylinder at the initial cross section defined by Eq. (3.8) at $\varphi = 1°$, and proceeds downstream to the point of separation of the laminar boundary layer (Fig. 3.1). An analytical definition for the boundary-layer separation is

$$\tau_w = \mu \left(\frac{\partial u}{\partial y} \right)_{y=0} = 0 \tag{3.16}$$

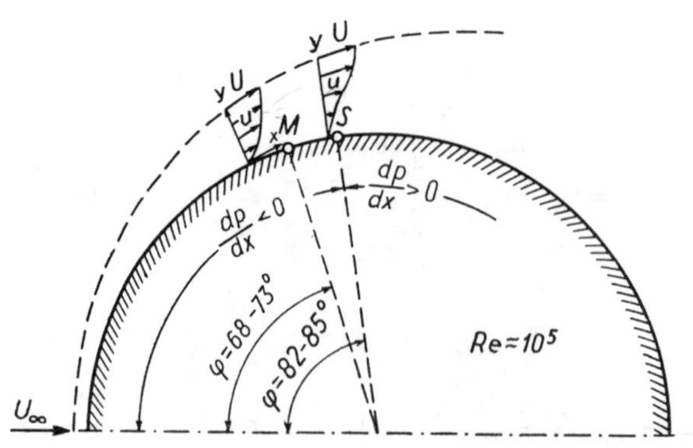

FIG 3.1 Fluid dynamics on a cylinder in the subcritical flow regime.

The separation point is characterized by zero transverse velocity gradient at the wall, and this point is downstream of that at which there is a minimum value of pressure [75]. The angle of separation calculated depends closely on the velocity distribution chosen for the outer boundary layer. Using Eq. (3.6), separation is calculated to occur at $\varphi = 82°$ and with a sinusoidal distribution of the potential velocity [53] of the form

$$\frac{U}{U_\infty} = 2 \sin \varphi \tag{3.6a}$$

Separation is calculated at $\varphi = 105°$.

A modification of the finite-difference technique of Patankar and Spalding [71] was applied in obtaining these solutions. In this modification [79], a partial derivative solution of the differential linear equations was substituted for the original finite-difference one. Our solution approaches the difference solutions found by Karyakin and Sharov [80] and by Jones and Launder [81].

3.3 THE EFFECTS OF BLOCKAGE FACTOR ON THE FLUID DYNAMICS AND HEAT TRANSFER

Most analytical and experimental data on single-cylinder heat transfer refer to infinite flow conditions. However, almost all practical applications leave significant blockage, which may exert an influence on the heat transfer.

Our experiments with different free-stream geometries yielded data for the effect of blockage factor k_q. The effect of blockage factor on velocity outside the boundary layer can be expressed in terms of a modified form of the Hiemenz relationship [Eq. (3.6)],

$$U/U_\infty = A_1^* \frac{x}{d} + A_2^* \left(\frac{x}{d}\right)^3 + A_3^* \left(\frac{x}{d}\right)^5 \tag{3.6b}$$

where

$$A_1^* = 3.6314 \left(1 + \frac{k_q}{2}\right) \tag{3.17}$$

$$A_2^* = -2.1709 \, (1 - 12.1284 \, k_q^{2.226} + 3.7376 \, k_q) \tag{3.18}$$

$$A_3^* = -1.5144 \, (1 + 18.542 \, k_q^{2.277} - 6.878 k_q) \tag{3.19}$$

At $k_q = 0$, Eq. (3.6*b*) coincides with the Hiemenz relation, Eq. (3.6). Figure 3.2 presents a comparison of Eq. (3.6*b*) with the experimental results. For $k_q = 0.3$, the velocity distribution approaches that given by Eq. (3.6*a*), as noted also by Akylbaev et al. [39].

To estimate the heat transfer behavior of a cylinder taking account of blockage, the same technique was used as for the cylinder in an effectively infinite flow, except that Eq. (3.6*b*) was introduced for free-stream velocity instead of Eq. (3.6). Figure 3.3 shows clearly that the location of the boundary-layer

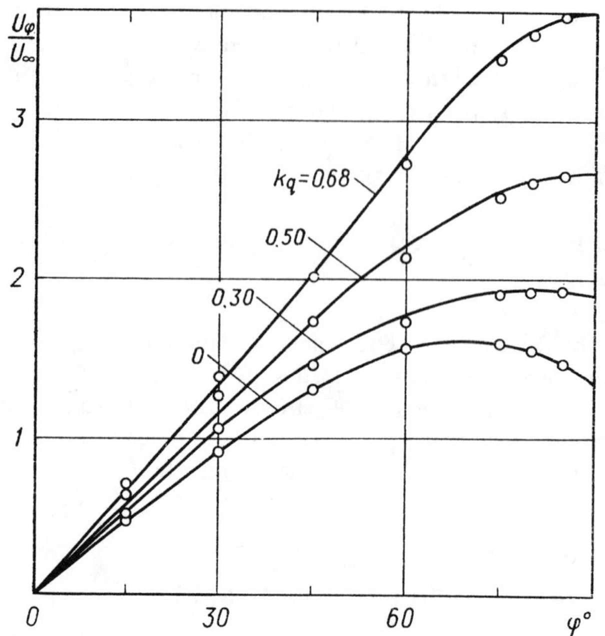

FIG. 3.2 Effect of blockage factor on the velocity distribution outside the boundary layer.

separation is related to blockage factor; this shift in position is caused by the change in the velocity distribution in the outer boundary layer.

3.4 PSEUDO-LAMINAR FLUID DYNAMICS AND HEAT TRANSFER AT THE FRONT PART OF A CYLINDER

At $2 \cdot 10^5 \leqslant \text{Re} \leqslant 10^7$, the heat transfer on the front part of a cylinder is affected strongly by free-stream turbulence, since the boundary layer is pene-

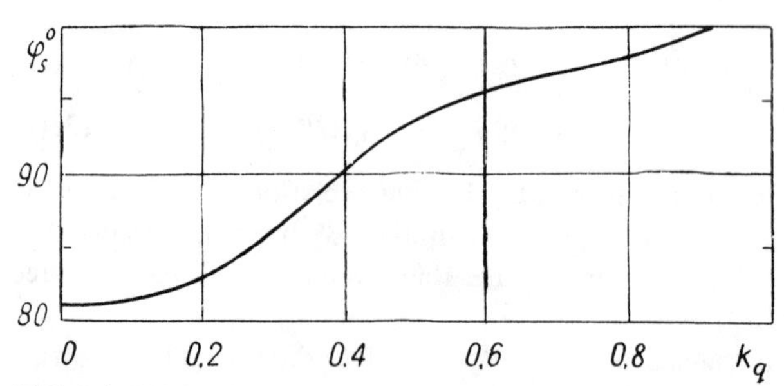

FIG. 3.3 Effect of blockage factor on the angular location of the separation point in the subcritical flow regime.

trated by velocity fluctuations from the free stream. In this range of Re, high turbulence is inevitably present in the test loops, so that the effect must be accounted for in the analytical techniques. For the lower range of Re, at least, the boundary layer can be treated as pseudo-laminar with enhanced turbulent viscosity and thermal conductivity due to the presence of turbulence [63]. This approach results in the following relations for the shear stress and heat flux:

$$\tau = (\mu + \mu_{Tu}) \frac{\partial u}{\partial y} \tag{3.20}$$

$$q = -(\lambda + \lambda_{Tu}) \frac{\partial T}{\partial y} \tag{3.21}$$

An analysis of our results, supported by the work of Roshko [27], by suggestions of Smith and Kuethe [59], and by McDonald and Kreskovsky [64], led to the choice of the following modified equation for turbulent viscosity

$$\mu_{Tu} = K \frac{Tu}{100} \delta U_{\infty} n^* \tag{3.22}$$

and correspondingly, for turbulent thermal conductivity

$$\frac{\lambda_{Tu}}{c_p} = \frac{\mu_{Tu}}{\mathrm{Pr}_t} \tag{3.23}$$

where $K \approx 0.15$ (an empirical constant), $\mathrm{Pr}_t = 0.8$ and n^* is a damping factor given by

$$n^* = [1 - \cos(\pi (y/\delta)^k)]/2 \tag{3.24}$$

For estimating μ_{Tu} from Eq. (3.22) for use in Eq. (3.20), k is taken as unity but for use in Eq. (3.23) k is taken as a function of Pr:

$$k = 2 \log \mathrm{Pr} + 1.37 \tag{3.25}$$

The use of an earlier equation [59]

$$\mu_{Tu} = 0.16\, y \frac{Tu}{100} \cdot U_{\infty} \tag{3.22a}$$

for μ_{Tu} was also investigated. It was shown that Eq. (3.22*a*) was applicable only for air; the discrepancy between the experimental results and the prediction using Eq. (3.22*a*) increases with Pr.

Calculations with $\mathrm{Re} < 10^5$ and *Tu* from 0 to 15% showed a downstream shift of the boundary layer separation point from $\varphi = 82$ to $\varphi = 88°$ (Fig. 3.4). The shift is a reflection of the change in τ with *Tu* in the range covered by Eqs. (3.20) and (3.21). The term μ_{Tu} in Eq. (3.20) is thus the third most important determining factor in the location of the separation point of a laminar or a pseudo-laminar boundary layer, ranking after those of velocity distribution in

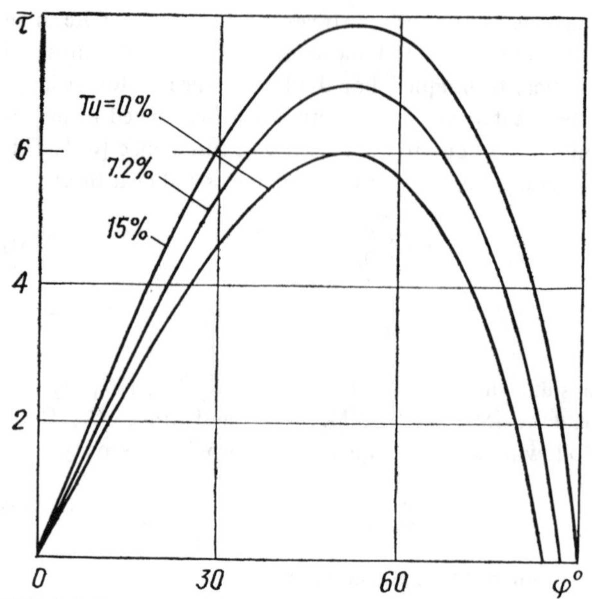

FIG. 3.4 The effect of free-stream turbulence on shear stress in the laminar boundary layer.

the outer boundary layer, and of blockage factor. The separation point (as indicated by the change in slope of the velocity curve, Fig. 3.5) is at $\varphi = 83.4°$ at $Tu = 1.7\%$. An increase of turbulence (to $Tu = 7.2\%$) is not accompanied by a deformation of the velocity profile. This suggests that the separation point is shifted downstream, and this view is confirmed by examination of the change in the shear stress profile (separation occurs when $\bar{\tau} = 0$, see Fig. 3.4).

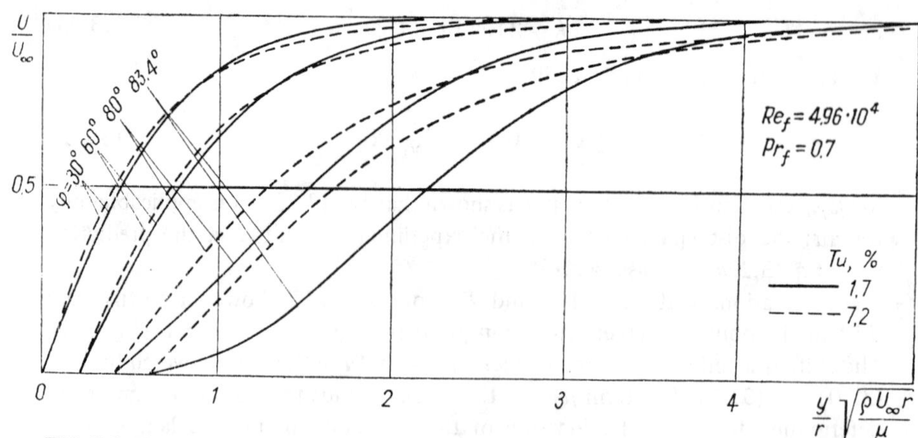

FIG. 3.5 The effect of free-stream turbulence on velocity profiles in the laminar boundary layer.

With the onset of the critical flow regime (Re > 10^5), the combined effect of Re and *Tu* is to shift the separation point to $\varphi = 90\text{–}100°$, i.e., a more significant shift than in the subcritical regime (Fig. 3.6). Experiments show [26] that the velocity distribution in the outer boundary layer is close to the potential velocity distribution of Eq. (3.6*a*) on the front part of a cylinder in the critical flow regime. Thus, with $k_q = 0$, neither Eq. (3.6) nor Eq. (3.6*b*) gives a proper description of the real velocity distribution at the edge of the boundary layer. They must be replaced either by Eq. (3.6*a*) or by a modification of Eq. (3.6) to

$$\frac{U}{U_\infty} = \left[3.6314 \frac{F_1 x}{d} - 2.1709\left(\frac{F_1 x}{d}\right)^3 - 1.5144\left(\frac{F_1 x}{d}\right)^5\right] \cdot F_2 \quad (3.6c)$$

where F_1 and F_2 are functions of Re and *Tu* (Table 3.1).

The inclusion of F_1 and F_2 in Eq. (3.6*c*) gives an improvement in the prediction of velocity U over a wide range of Re, and a better agreement with the experimental heat transfer results. This modification also allows the more exact location of the separation point, where $\tau_w = \mu(\partial u/\partial y)_{y=0} = 0$.

For a pseudo-laminar boundary layer on a cylinder ($Tu \geqslant 1\%$) in a supercritical crossflow, a similar technique of calculation is applied as for the subcritical region, but Eq. (3.6*c*) is used instead of Eq. (3.6) or (3.6*a*). Also, in the supercritical region, T_w cannot be described by Eq. (3.15); however, an alternative relation for the front stagnation point has been suggested [38, 74] as follows:

$$\mathrm{Nu}_{fd} = 0.326\, \mathrm{Re}_{fd}^{0.6}\, \mathrm{Pr}_f^{0.33}\, Tu^{0.15}\, (\mathrm{Pr}_f/\mathrm{Pr}_w)^{0.25} \quad (3.15a)$$

where $\mu, \lambda, \rho, c_p = f(t)$ by the analogy with [79].

Figure 3.7 gives sample temperature profiles and illustrates the deformations

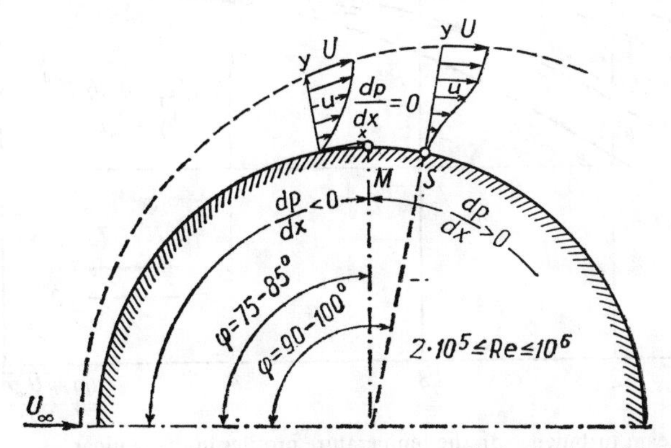

FIG. 3.6 Flow dynamics for flow over a cylinder in the critical regime.

TABLE 3.1 Values of the Functions F_1 and F_2 in Eq. (3.6c) for $k_q = 0.16–0.25$

	Re = 10^3		Re = 10^4		Re = $2 \cdot 10^5$		Re = 10^6	
Tu (%)	F_1	F_2	F_1	F_2	F_1	F_2	*F1*	F_2
0.25	1	1.11	1	1.11	0.8	1.10	0.86	1.02
1.00	1	1.13	1	1.12	0.8	1.12	0.86	1.04
2.50	1	1.14	1	1.18	0.8	1.28	0.86	1.18
5.00	1	1.16	1	1.19	0.8	1.26	0.86	1.19
10.00	1	1.18	1	1.22	0.8	1.16	0.86	1.19
15.00	1	1.18	1	1.22	0.8	1.12	0.86	1.19

3.5 HEAT TRANSFER AND FLUID DYNAMICS IN THE TRANSITIONAL AND TURBULENT BOUNDARY LAYERS ON A CYLINDER

In the supercritical flow regime, considerable changes occur in the fluid dynamics of flow around a cylinder. The increase of Re results in a shorter laminar section of the boundary layer. The transition to a nonlaminar boundary layer is shifted from $\varphi \approx 100°$ to $\varphi \approx 30°$. The transition is followed by transitional and then turbulent regions and then by the separation of the turbulent boundary layer [24-26, 28]. This complex fluid dynamic pattern dictates the use of special techniques for heat transfer prediction.

The laminar part of the boundary layer at the front of the cylinder may be treated by the methods described previously; however, it is necessary to locate the transition point.

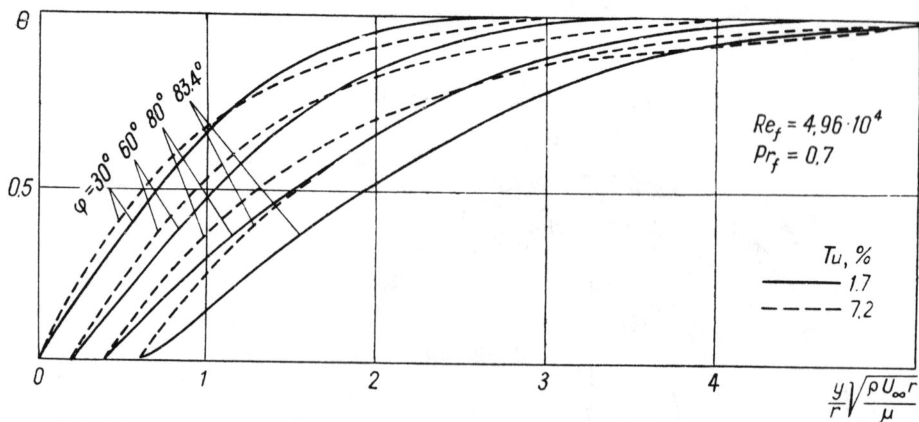

FIG. 3.7 The effect of free-stream turbulence on the temperature profiles in the laminar boundary layer.

The transitional boundary layer In this part of the boundary layer there is an adverse longitudinal pressure gradient, which tends to stabilize the layer [53]. Cebeci and Smith [67] suggested several techniques for determination of the start of the transitional boundary layer. However, the technique suggested by Michel [76], based on the universal relation between $Re_{\delta_p^{**}}$ and Re_{x_p}, is perhaps more appropriate. Cebeci et al. [77] suggested an analytical expression between $Re_{\delta_p^{**}}$ and Re_{x_p} and found a good agreement with experimental data for Re from 10^5 to $4 \cdot 10^7$:

$$Re_{\delta_p^{**}} = 1.174\,[1 + (22400/Re_{x_p})]\,Re_{x_p}^{0.46} \tag{3.26}$$

where $Re_{\sigma_p^{**}} = U\delta^{**}\rho/\mu$, $Re_x = Ux\rho/\mu$.

In Eq. (3.26) the value of x refers not to the generatrix on the cylinder, but to the start of the pseudo-laminar boundary layer, this location being determined by the momentum balance equation [79]:

$$\tau_w\,dx = d\left\{\int_0^\infty \rho u\,(U - u)\,dy\right\} \tag{3.27}$$

It follows from Eqs. (3.26) and (3.27) that the location of the transitional boundary layer is governed by the velocity profile–that is, by the boundary layer thickness and by the values of Tu, Re, and dp/dx, because both the value of x from Eq. (3.27), and the value of

$$\delta^{**} = \int_0^\infty \frac{u}{U}\left(1 - \frac{u}{U}\right)dy \tag{3.28}$$

are evaluated by integration from equations that contain integrals of the velocity profile. Experimental and analytical results suggest that the point of onset of the transition boundary layer shifts upstream with an increase of turbulence. In this case, an increase of free-stream turbulence leads to the instability of the pseudo-laminar boundary layer, and to an earlier onset of the transitional flow regime. The beginning of the transitional boundary-layer region is determined in the following way: for each value of x, in the downstream direction, solutions of the pseudo-laminar boundary layer equation are obtained, and the difference values

$$|\,Re_{\delta^{**}}\,| - |\,1.174\,[1 + (22400/Re_x)]\,Re_x^{0.46}\,| \tag{3.29}$$

are determined. The onset of the transition layer is located at the minimum value or a zero value of the difference. Equation (3.26) is derived from experimental results on cylinders ; however, we have obtained an alternative relation using numerical analysis. This alternative relationship–which can be used instead of Eq. (3.26)–is as follows:

$$Re_{\delta_p^{**}} = 1.047\,[1 + (22400/Re_{x_p})]\,Re_{x_p}^{0.46} \tag{3.26a}$$

In real fluids, the laminar–turbulent transition is a gradual process in the boundary layer, and covers a finite region on the surface. This region is characterized by the intermittence factor, which is $\gamma = 0$ for laminar flows, and $\gamma = 1$ for turbulent flows. From reference 70, for a flat boundary layer, we have

$$\gamma = 1 - \exp[-(G/U)(x - x_p)^2] \tag{3.30}$$

where $G = (3/C^2)(U^3/\nu^2)\,\mathrm{Re}_{xp}^{-1.34}$, $C = 60 + 4.86M_f^{1.92}$, $0 < M_f < 5$, $\mathrm{Re}_{xp} = Ux_p/\nu$.

The turbulent boundary layer In the general formulation, the values of τ and q are

$$\tau = (\mu + \mu_{Tu} + \mu_t\gamma)\frac{\partial u}{\partial y} \tag{3.31}$$

$$q = -(\lambda + \lambda_{Tu} + \lambda_t\gamma)\frac{\partial T}{\partial y} \tag{3.32}$$

In the solution, the first approximation [71] involves making the following common assumptions about the turbulent transport

$$\mu_t = l^2\left|\frac{\partial u}{\partial y}\right| \tag{3.33}$$

$$\frac{\lambda_t}{c_p} = \frac{l^2}{\mathrm{Pr}_t}\left|\frac{\partial u}{\partial y}\right| \tag{3.34}$$

where

$$l = \varkappa y \quad \text{when} \quad 0 < y \leqslant \alpha^*\delta/\varkappa \tag{3.35}$$

$$l = \alpha^*\delta \quad \text{when} \quad \alpha^*\delta/\varkappa < y \tag{3.36}$$

Here α^* stands for the dependence of the turbulent transport on the Reynolds number

$$\alpha^* = \alpha_1^*\frac{1.55}{1+\Pi} \tag{3.37}$$

where $\alpha_1^* = 0.09$;

$$\Pi = 0.55[1 - \exp(-\sqrt{0.243z_1} - 0.298z_1)] \tag{3.38}$$

Here $z_1 = (\mathrm{Re}_{\delta^{**}}/425) - 1$ [78].

As the wall is approached, the turbulent transport is reduced to zero. To represent the damping out of μ_t and λ_t, a modification [65] of the van Driest damping factor n_1^* was introduced, which accounts for the effect of the longitudinal pressure gradient:

$$n_1^* = 1 - \exp(-y/A) \tag{3.39}$$

where

$$A = \frac{A_1}{\sqrt{1+K_1 \mu_t (dp/dx) (\tau_w^{*} \rho_f)^{-1/2}}} \tag{3.40}$$

$$A_1 = 26 + \frac{14}{[1+(\mathrm{Re}_{\delta^{**}}/1000)^2]} \quad \text{when} \quad \mathrm{Re}_{\delta^{**}} > 300 \tag{3.41}$$

The value $K_1 = 11.8$ is widely used in the literature. For an incompressible flat boundary layer K_1 can be interpreted as the value of the dimensionless distance parameter $y^+ = yu_*/\nu$ at the point of intersection of viscous sublayer velocity profile ν with the logarithmic profile, both profiles being commonly expressed in terms of the dimensionless velocity parameter $u^+ = u/u_*$.

After substitution from Eq. (3.39), the resultant expression for the mixing length is

$$l = n_1^* \varkappa y \tag{3.42}$$

According to [66], the universal von Karman constant is a function of $\mathrm{Re}_{\delta^{**}}$:

$$\varkappa = 0.40 + \frac{0.19}{1+0.49\left(\frac{\mathrm{Re}_{\delta^{**}}}{1000}\right)^2} \tag{3.43}$$

In the critical flow regime, Eq. (3.6*c*) is applicable for the vicinity of the front stagnation point and up to $\varphi = 90$–$100°$. We assumed in our analysis that $\kappa = 0.41 =$ constant.

In order to increase the level of accuracy, the value of dp/dx from Eqs. (3.6*a*) and (3.7) was derived from

$$f_3 = 1 + \frac{\varphi - 100}{100} \tag{3.44}$$

where $\varphi \geqslant 100°$. Juding only by the experiments, it can be assumed that $dp/dx = 0$ for $\varphi \geqslant 120°$. Thus, the experimental accuracy is by no means sufficient for the prediction of the pressure distribution for φ up to 140°, that is, up to the separation of the boundary layer.

The effect of surface curvature on the turbulent transport The existence of a convex or a concave surface on a body can reduce or augment heat transfer to the body. For a cylinder, Bradshaw [68] and Cebesi [69] suggested the following expression to take account of curvature effects on the turbulent transport:

$$S = \left(\frac{1}{1+7\,\mathrm{Ri}}\right)^2 \tag{3.45}$$

where Ri is analogous to the Richardson number, as in

$$\mathrm{Ri} = \frac{2U}{r}\left(\frac{\partial u}{\partial y}\right)^{-1} \tag{3.46}$$

The component terms of the turbulent transport μ_t and λ_t were multiplied by s. In the case considered, the effect of s is below 1% at the wall, but it reaches 15% near the edge of the outer boundary layer. For the laminar part of the boundary layer, the effect of surface curvature is negligible, because $r \gg \delta$.

3.6 NOTES ON THE IMPLEMENTATION OF THE ANALYTICAL TECHNIQUES

General notes The growth of the boundary layer introduces a high degree of complexity in prediction. The boundary-layer equations were solved in the coordinate system

$$\xi = x, \quad \omega = \frac{\psi - \psi_w}{\psi_f - \psi_w} \tag{3.47}$$

Here $d\psi = \rho u\, dy$ is a stream function [71]. The subscripts f and w on the stream function denote its values in a nondisturbed flow and at the wall, respectively. The growth of the boundary-layer thickness is related to the mass flow rate through the outer edge of the boundary layer $(\rho v)_f$. For the laminar and pseudo-laminar parts of the boundary layer, this value is determined from the continuity equation:

$$(\rho v)_f = -\left| \int_0^\infty \frac{\partial(\rho u)}{\partial x}\, dy \cdot (\psi_f - \psi_w)\left(\frac{U_j}{U_{j-1}} - 1\right)\left(\frac{1+\sin\varphi}{2}\right) \Big/ \Delta x \right| \tag{3.48}$$

where $j-1$ and j are two adjacent cross-section areas in the boundary layer normal to the tangent to the cylinder surface. Our expression for $(\rho v)_f$ in Eq. (3.48) was the result of a numerical study based on a comparison of analytical and experimental results.

As noted by Goldstein [83], the thickness of the boundary layer undergoes continuous growth over the transition region, reaching at the end of this region almost double its value at the beginning. To account for this, a special factor $K \approx 4$ was introduced in Eq. (3.48).

For the turbulent zone of the boundary layer, the expressions for the mass transfer rate are very complex. Wide experience has been accumulated at the Institute of Physical and Technical Problems of Energetics, Academy of Sciences of the Lithuanian SSR, on the various analytical techniques and their behavior. However, so far, we have not been able to adduce a solution for the mass transfer through the outer edge of the boundary layer. Comparisons of the numerical results and the various experimental implementations have shown a certain amount of disagreement. We can only suggest several reasons for the difficulties encountered:

1. The poor performance of the model [Eq. (3.30)] for the laminar–turbulent transition zone, probably because it had been primarily intended for flat plates and aerofoils, rather than for circular cylinders.
2. The highly complicated fluid dynamics at high velocities of free-stream turbulence, where insufficient account has been taken of the large longitudinal pressure gradients and of their effects on the velocity and temperature profiles.
3. The complex manner of formation and thickness growth of the boundary layer.
4. Difficulties in describing the velocity variation at the edge of the turbulent boundary layer.

In this study, we have ignored the case of the laminar–turbulent transition immediately following separation of the laminar boundary layer and formation of the separation bubble, since the latter is not large in size, occupies an angle $<10^\circ$, and is only observed in a limited range of Re.

Choice of the integration step For a successive integration along x (that is, along the circumference), the step length of integration must not exceed one-third of the laminar boundary layer thickness,

$$\Delta x \leqslant 0.3\,\delta \tag{3.49}$$

The presence of the nondimensional transverse coordinate in Eq. (3.47), suggested by Patankar and Spalding [71], allows restriction of the number of points required to cover the full range of x to:

$$N = \text{constant} \approx 100$$

For the laminar and pseudo-laminar parts in the subcritical and critical flow regimes the choice of step length is arbitrary. At higher Re, for both the

TABLE 3.2 Values of c and K as Functions of Re

	Re		
	10^4–10^5	10^5–10^6	10^6–10^7
c	5	7	10
K	1.04	1.05	1.06

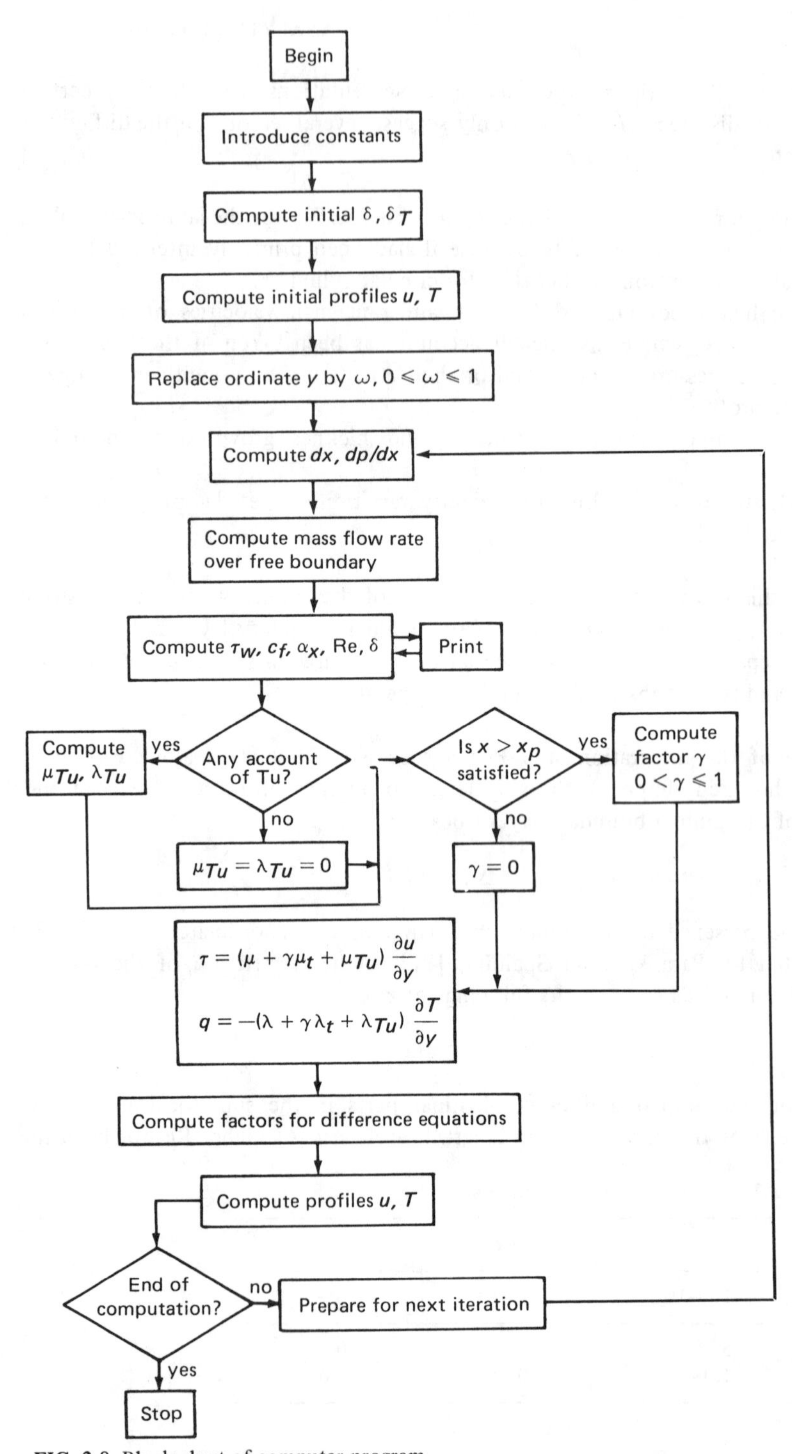

FIG. 3.8 Block chart of computer program.

transitional and the turbulent zones, the suggestions given in references 80, 81 on the choice of the integration step should be borne in mind. After a preliminary analysis to determine the optimum, the following distribution was assumed:

$$y_2 = \frac{\delta}{N \cdot c} \tag{3.50}$$

$$y_i = y_{i-1} + G_{i-1}, \quad i = 3,\ 4,\ \ldots,\ N+1 \tag{3.51}$$

where $N \approx 100$, $G_i = G_{i-1} \cdot K$, and G_2 is given by the geometrical progression

$$G_2 = \frac{(\delta - y_2)(K-1)}{K^{(N+1)} - 1} \tag{3.52}$$

The values of K and c are given in Table 3.2. A block-chart of the computer program is presented in Fig. 3.8.

CHAPTER
FOUR

SPECIFIC FEATURES OF FLUID DYNAMICS OF CROSSFLOW OVER A CYLINDER

There exists a close relation between the dynamics of crossflow over a cylinder on the one hand, and the heat transfer between the fluid and the cylinder on the other. Sophisticated studies of such dynamic processes as pressure distribution, variation of free-stream velocity, and friction drag are needed to achieve a proper understanding of these relations. Here, great importance must also be ascribed to the effect of free-stream turbulence, and to the effect of shell-channel blockage. Each of these processes influences in some way or other the fluid dynamics on the surface, the separation of the boundary layer, and formation of the near wake. In our program, all these effects were studied experimentally; in the present section we deal with the fluid dynamics of crossflow of air, water, and transformer oil over circular and elliptic cylinders. The tests covered wide ranges of Reynolds and Prandtl number.

4.1 A COMPARISON OF IDEAL AND REAL FLUID DYNAMICS OVER A CYLINDER

Cylinder in Crossflow of an Ideal Fluid

An analysis using ideal or potential fluid dynamics, although it ignores internal friction or viscosity, nevertheless yields in a simplified manner the main flow parameters–i.e., the velocity distribution and the pressure distribution on the surface.

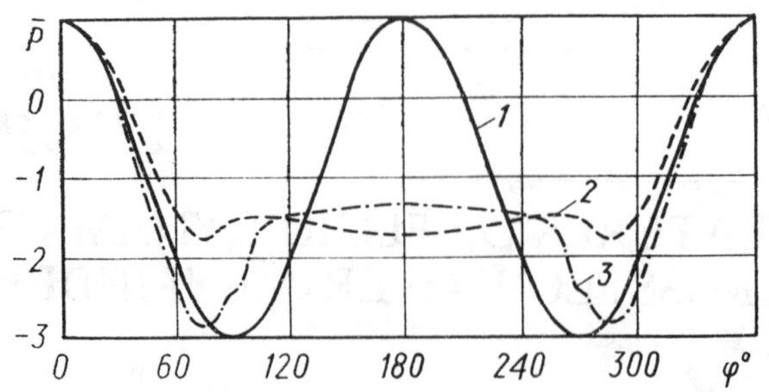

FIG. 4.1 Pressure distribution on a cylinder surface: 1, potential flow (analysis); 2, subcritical flow of air (Re = $8 \cdot 10^4$) (experiment); 3, critical flow of water (Re = $2 \cdot 10^5$, $k_q = 0.5$) (experiment).

Figure 4.1 presents pressure distribution curves obtained from the potential flow solutions. They are compared to the experimental results observed in different flow regimes. The agreement is satisfactory only for the front part of the cylinder.

The velocity distribution from ideal fluid dynamics–that is, in the absence of the boundary layer–is described by

$$U_\varphi = U_\infty \sin \varphi \left[1 + \left(\frac{r}{r_1}\right)^2\right] \tag{4.1}$$

where r and r_1 are the axial distances to the cylinder surface and to the test point, respectively. Equation (4.1) suggests that lower velocities are obtained at larger axial distances. The fluid velocity on the surface is

$$U_\varphi = 2U_\infty \sin \varphi \tag{4.2}$$

Thus at the front stagnation point ($\varphi = 0$), the tangential velocity U_φ is at its minimum and is equal to zero. It increases with the angular distance and, in the central cross section ($\varphi = 90°$) reaches double the free-stream value. Substitution of Eq. (4.2) into the Bernoulli equation gives a functional relation between flow velocity and pressure in the flow:

$$\bar{p} = \frac{2\,(p_\varphi - p_\infty)}{\rho U_\infty^2} = 1 - 4 \sin^2 \varphi \tag{4.3}$$

Equation (4.3) suggests an inverse dependence between the pressure and the velocity. It is seen from Fig. 4.1 that the pressure is at a maximum at the stagnation point, and at a minimum at the central section. For an ideal fluid, the pressure would be distributed symmetrically and integration of of the pressure distribution results in a zero value for the force on the cylinder. The ideal fluid case is thus different from that for real fluids where the pressure distributions are nonsymmetrical (Fig. 4.1). The nonsymmetrical pressure distribution results in a net force on the cylinder, and the existence of this force is the main cause of the pressure drop across the cylinder. The pressure distribution calculated from potential flow applies only in the front part of a cylinder for flows of real fluids.

Flows of Real Fluids

The ideal model shows some specific features of fluid dynamics on a cylinder, but it never applies directly to a real fluid because of the influence of viscosity.

The viscous interaction of the fluid and the tube surface gives rise to the boundary layer, as discussed in Chap. 3. With a cylinder in crossflow, a laminar boundary layer is formed on the upstream part, its thickness increasing downstream. The factors that determine the fluid dynamics in the boundary layer are the Reynolds number and the free-stream turbulence. Curves 2 and 3 in Fig. 4.1 are drawn through the measured values of pressure in flows of water and air; included are data showing the effect of the blockage factor and the turbulence. In real fluids, the relative importance of inertia and viscous forces is governed by the Reynolds number; Re may be used as a parameter to delineate the various flow regimes.

For $Re < 1$, inertia is overshadowed by viscosity, and the flow envelopes the tube in laminar streams, which separate only in the rear stagnation point (Fig. 1.1).

For $3 < Re < 5$, the streams begin to deviate in the rear part of the cylinder, and the laminar boundary layer separates from the surface. Two symmetrical steady-state vortices are formed in the wake. The vortices are bound by streams of zero velocity, which are prolongations of the laminar boundary layer after its separation. A laminar steady-state flow is formed in the wake. With a further increase of Re (>40), the wake becomes unstable and vortex shedding is initiated. First, one of the two steady-state vortices separates from the tube, and then the second is shed because of the nonsymmetric pressure in the wake. The intermittently shed vortices form a wake called the von Karman vortex street, which is also steady-state and laminar over a substantial distance behind a cylinder.

In the range $150 < Re < 300$, periodic irregular disturbances are observed in the wake. In this range the flow is transitional, and gradually becomes a turbulent one (with a vortical wake) as Re increases. A three-dimensional structure has been observed in this flow range [84]. With a further increase of Re, the wake becomes turbulent in the region of vortex formation.

Turbulent vortex flow is observed in the wake up to the establishment of the critical flow regime at $Re \approx 2 \cdot 10^5$, this transition being characterized by a sharp decrease in the pressure drag, a higher rarefaction at the back of the cylinder, and irregular vortex shedding. The irregular vortex shedding is normally considered to continue up to the limit of the critical flow, which is at $Re = 6 \cdot 10^5$. However, recent work by Roshko [27] shows that regular vortex shedding *does* exist in the critical flow region and, indeed, beyond it (i.e., at $Re > 3.5 \cdot 10^6$).

4.2 BOUNDARY-LAYER SEPARATION

The separation of the boundary layer from the cylinder surface is a determining factor for a number of processes: pressure drop, velocity distribution near the surface and in the external layers, flow in the wake, flow-induced vibration, etc. Knowing the manner of separation and of its location is highly important.

The separation phenomenon is governed by friction and by changes in velocity and pressure. In the boundary layer on the front part of the cylinder, the energy of compression of the fluid is transformed to kinetic energy. A reverse transformation occurs in the rear. The variations of the flow velocity and pressure are described by the Bernoulli equation:

$$p+\frac{\rho U_{\infty}^{2}}{2}=\text{const} \tag{3.7}$$

On the front part, the pressure decreases downstream ($dp/dx < 0$), and velocity increases accordingly. Fluid particles in the boundary layer are entrained by the external flow and continue their forward movement near the cylinder surface, in spite of friction.

On the rear part of the cylinder, the pressure increases ($dp/dx > 0$) and the velocity decreases downstream. In the region outside the boundary layer, the pressure increase is reflected in a decrease in velocity but forward motion of the fluid continues. However, in the boundary itself, where viscous effects slow the fluid, the increase in local pressure (caused primarily by the flow outside the layer) causes a rapid deceleration of the fluid to rest, and finally a flow reversal is observed in the boundary layer. The reverse streams are curled up to form a vortex, and separate the surface. In the subcritical flow regime, the separation is observed at $\varphi = 80^{\circ}$ (Fig. 4.1, curve 2). This observation was supported by our experimental studies of the heat transfer in different fluids at wide ranges of Re.

We conclude that a positive pressure gradient ($dp/dx > 0$) and the presence of viscous forces in the boundary layer are two necessary conditions for boundary layer separation. For details on the analytical determination of the boundary-layer separation, see Section 3.2.

In our studies, particular attention was paid to the critical range of Re. The results in flows of water and air suggest a sudden forward shift of the separation point to $\varphi = 140^{\circ}$. Achenbach [26] reported a further backward shift to $\varphi = 120^{\circ}$ for the separation point at $\text{Re} = 10^{6}$. This might be an indication of the new, supercritical flow regime. However, this phenomenon was not observed in our studies with water and air (Fig. 4.2).

The blockage factor was found to have little effect on the behavior of the separation point. In the subcritical regime, the separation point was shifted from $\varphi = 80^{\circ}$ to $\varphi = 95^{\circ}$ with the variation of k_q from 0.25 to 0.83 [85]. In the critical flow regime, a 10° upstream shift of the separation point was noted with an increase of the blockage factor.

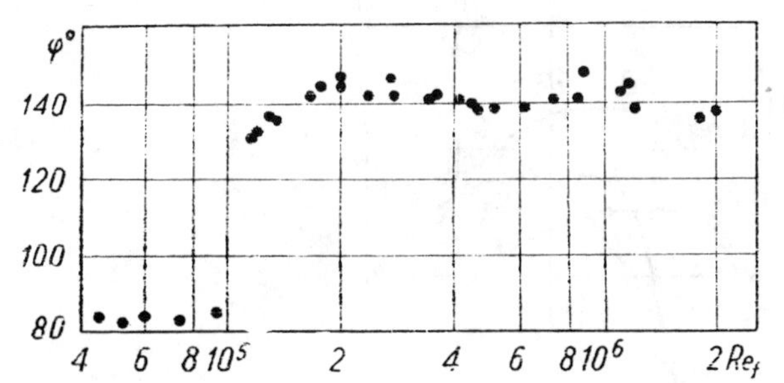

FIG. 4.2 Dynamic behavior of boundary-layer separation.

A specific phenomenon related to the fluid flow on cylinders and aerofoils is the separation bubble, mentioned in a previous section [27, 28]. It is formed by a partial separation of the laminar boundary layer, followed by its reattachment and by the separation as a turbulent layer. The separation bubble is formed under the influence of the external flow layers and occurs in a specific range of Re (Fig. 4.1, curve 1). The flow pattern associated with a separation bubble is shown in Fig. 4.3.

The successive phenomena related to the fluid dynamics on a cylinder are reflected in the behavior of the base pressure value $p_{180°}$, i.e., the pressure in the rear critical point, with Re. The results of our analysis are presented in Fig. 4.4. Analogous results in air were reported by Roshko [86]. Processes related to the separation bubble are reflected by part ABC of the curve of the base pressure variation. For the critical flow regime, AB corresponds to a growing separation bubble, reaching a maximum size at B and decaying in the region BC. Here the boundary-layer separation is preceded by laminar-turbulent transition. The qualitative agreement of the points in Fig. 4.4 is satisfactory, but for larger disturbances and blockage factors (Tu and k_q) the separation bubble can be initiated at lower values of Re_f.

The concept of the laminar-turbulent transition of the boundary layer is at present interpreted in two different ways. The first theory, proposed by a number of authors [27, 28, 86, 103], locates the transition after the separation

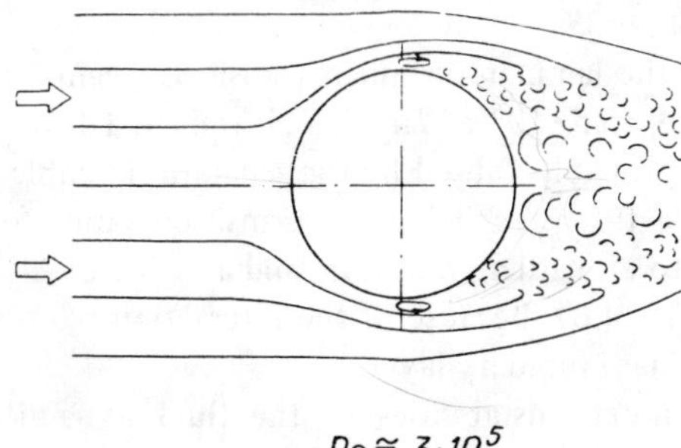

FIG. 4.3 A schematic representation of the flow patterns associated with a separation bubble.

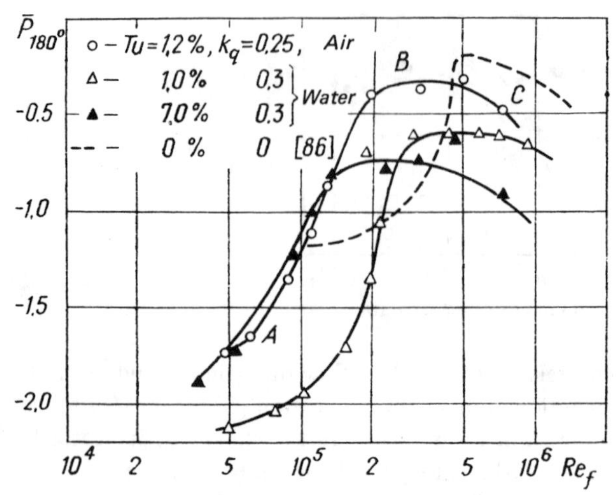

FIG. 4.4 Variation of the base pressure with Re.

bubble, so that at the critical values of Re, the laminar boundary-layer separation on the front part of the cylinder is followed by a separation bubble, a reattachment, and a turbulent boundary layer. The second theory is based on the heat transfer observations [31] and says that at the critical values of Re, the laminar boundary layer loses its stability, so that a transitional part of the boundary layer occurs followed by a turbulent part. Both the separation and the transition occur at about $\varphi = 140°$. The second interpretation would obviously apply in the absence of the separation bubble. We suggest the first interpretation is applicable to the critical flow regime, since it is consistent with the measurements of the shear stress and of the pressure distribution reported below. The second interpretation applies to the supercritical flow regime.

Our measurements show that the separation bubble occupies a 10° angular region on the cylinder. The laminar separation and the laminar–turbulent transition are very close together, so that with sufficient accuracy, we may locate the beginning of the transition at the point of minimum heat transfer. Figures 4.5 and 4.6 illustrate the variation of the transition point with Re_f, turbulence level, and blockage factor for water and air [24, 37].

For Re_f from $2 \cdot 10^5$ to $3 \cdot 10^5$, the laminar–turbulent transition begins at $\varphi = 80$-$85°$, but with an increase of Re_f to $5 \cdot 10^5$ at $Tu \approx 1\%$, it is moved downstream to $\varphi = 95°$. The shift is most probably caused by the separation bubble. A further increase of Re_f leads to a further shift of the transition point to $\varphi = 35°$. Increasing turbulence leads to an earlier transition, and a higher blockage factor has a stabilizing effect (Fig. 4.6), because of the interaction of the higher pressure gradients with the laminar boundary layer.

Some observers [87-90] have noted instabilities of the fluid dynamic parameters. Velocity fluctuations have been observed near the front stagnation

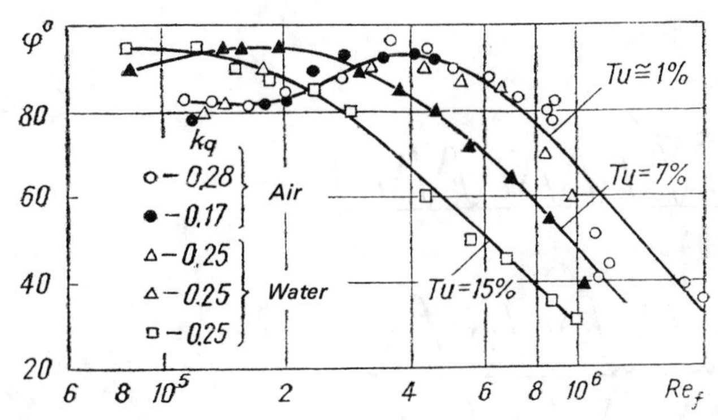

FIG. 4.5 Dynamic behavior of the onset of the laminar–turbulent transition in the boundary layer.

point. Pressure shocks and vortex shedding at two (rather than one) characteristic frequencies have also been reported. In view of these observations, we included in our program a separate study, in subcritical air flow, a specially constructed tube with $d = 70$ mm [91]. Anemometer measurements of the average velocity and of the velocity fluctuations were performed on the circumference and at different radial distances from the tube surface.

Any changes in the location of the front stagnation point should be reflected by changes of flow direction, the zero-velocity line crossing a specified point in the space. In Fig. 4.7, the left-hand curves present the actual velocity fluctuations at three separate points near the cylinder surface. The right-hand curves are the corresponding oscillograms for $\mathrm{Re}_f = 4 \cdot 10^4$, $\varphi = 0$, 4, and $-4°$, at a 0.95-mm radial distance from the surface. The measurement was performed synchronically in the three points; the signals from the anemometers are clearly nonsinusoidal. This nonsinusoidality may be the cause of the *apparent* existence of two vortex-shedding frequencies near the front and rear stagnation points.

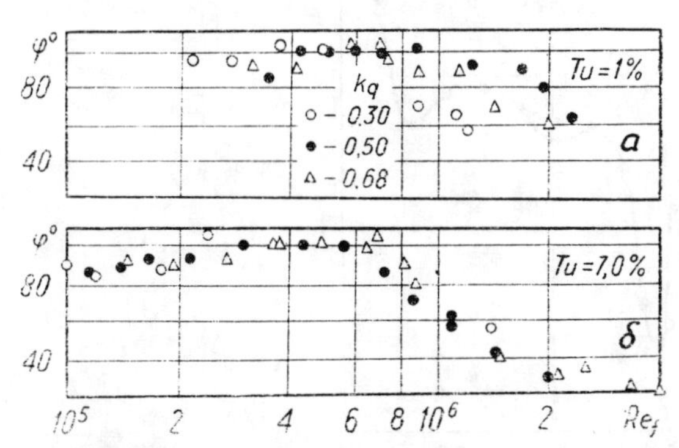

FIG. 4.6 The effect of the blockage factor k_q on the onset of the laminar–turbulent transition in the boundary layer.

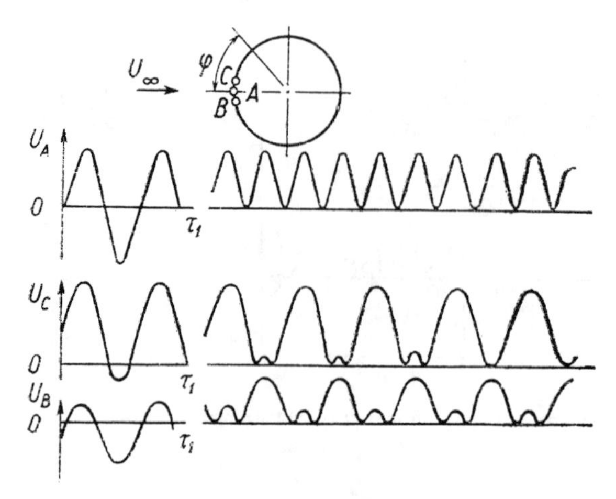

FIG. 4.7 Velocity fluctuation curves in the vicinity of the front stagnation point.

If the double frequency is real, it should be observed in the whole cycle of the fluctuation amplitude. Reliable information on the subject can only be read from linear correlation curves of velocity if the flow is quasi-steady (Fig. 4.8). The curve was recorded with the sensors located symmetrically around the circumference with respect to the front stagnation point so that both the single and the double vortex (Strouhal) frequencies could be indicated. The particular

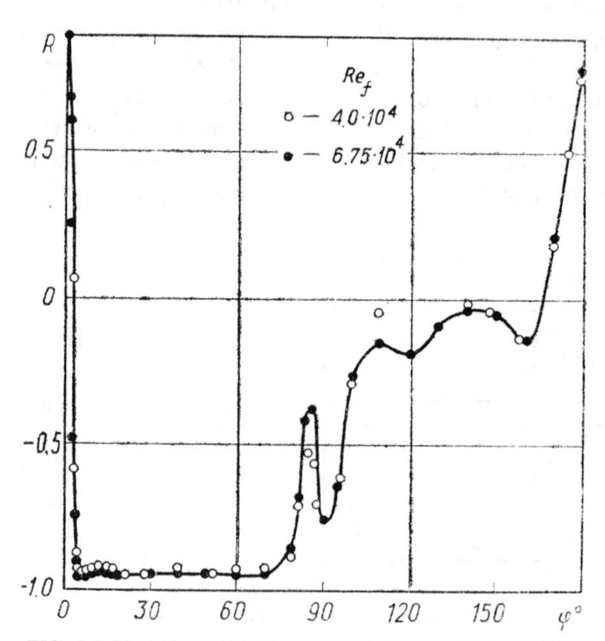

FIG. 4.8 Variation of the linear correlation coefficient with φ.

record refers to $Re_f = 4 \cdot 10^4$ and $6.75 \cdot 10^4$ at 1.02 mm from the surface. Changes of Re_f or of the radial distance had no qualitative effect on the shape of the curves.

The dynamic behavior of the stagnation points and of the separation points is demonstrated in Fig. 4.9 using synchronous velocity oscillograms for $\varphi = +3$, $+95$, -95, and $+175°$ taken at $Re_f = 2 \cdot 10^4$ and at 1.5 mm over the surface. When the stagnation point, as indicated by a minimum velocity, is shifted in the clockwise direction ($\varphi = +3°$), the subsequent curves ($\varphi = +95$ and $-95°$) show a corresponding upstream shift of the upper separation point (minimum velocity and turbulent flow), and a simultaneous downstream shift of the lower separation point. A synchronous motion of the rear stagnation point is also evident. Thus a vortex formed on one side of the cylinder causes the opposite separation point to move upstream.

Our measurements of the fluctuation ranges covered by the specific points gave $\varphi = \pm 6°$ for the front stagnation or impact point, and $\varphi = \pm 15°$ for the rear stagnation point.

Anemometer signals for the separation points ($\varphi = +95$ and $-95°$), shown in Fig. 4.9, reflect clearly the intermittent structure of the flow. When a separation point is downstream of the sensor, the latter indicates a laminar maximum velocity flow. Whenever a separation point is shifted upstream, the sensor indicates a separated vortex, minimum-velocity, flow.

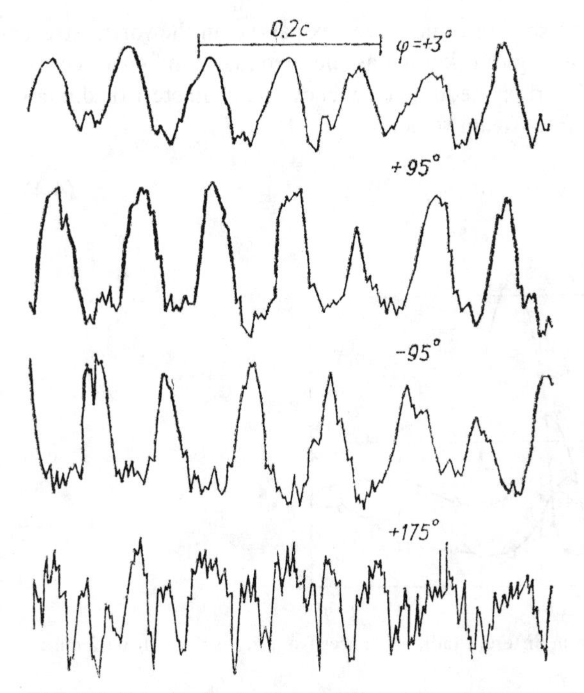

FIG. 4.9 Velocity fluctuation in different locations on a cylinder.

A similar picture of the dynamics of the separation point and of the separated stream is given by the results for mean-square fluctuation of free-stream velocity (Fig. 4.10). The sharp peaks in the curves for higher distances from the surface (curves 2 and 3) against a flat maximum on the wall (curve 1) reflect the sharp increase in the amplitude of velocity fluctuations in the separated stream.

The measurements suggested that all the fluid dynamic parameters for flow over a cylinder undergo fluctuations at the Strouhal frequency. The double frequencies indicated by thermoanemometers in the stagnation (front and rear) regions result from the fact that hot wire sensors can give only modulus signals. The fluctuation ranges are $\varphi = \pm 6°$ for the front stagnation point, $\varphi = \pm 15°$ for the rear stagnation point, and $\varphi = \pm 20$ to $25°$ for $\varphi = 80$–$105°$ for the separation points. With each vortex shed, the stagnation points perform a shift in one direction, opposite to that of the front stagnation and the separation points.

4.3 VORTEX FLOW IN THE WAKE

The wake is formed by the separated boundary layers and the vortices, which either circulate in a stationary position or are regularly shed from the cylinder and form the so-called vortex street. Their behavior is governed by the Reynolds number, the free-stream turbulence, the blockage factor, and the nature of the surface.

The frequency of vortex shedding and of vortex passage in the vortex street is described by a dimensionless group known as the Strouhal number, given by the ratio of the product of vortex shedding frequency and characteristic dimension (e.g., diameter), to the free-stream velocity.

$$\mathrm{Sh} = \frac{fd}{U_\infty} \tag{4.4}$$

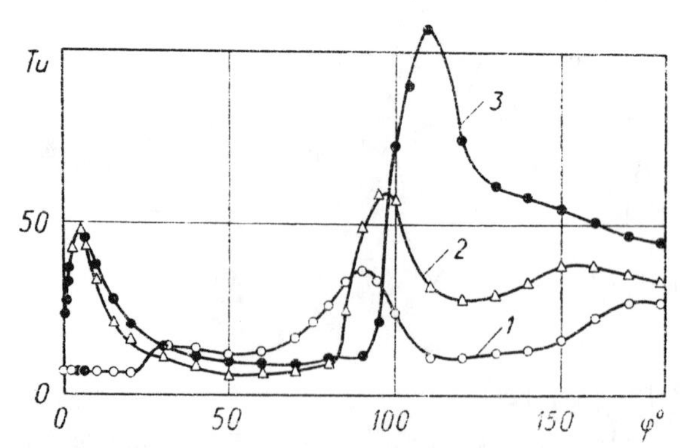

FIG. 4.10 Velocity fluctuation in different radial distances from the surface: 1, 0.02 mm; 2, 1.02 mm; 3, 3.0 mm.

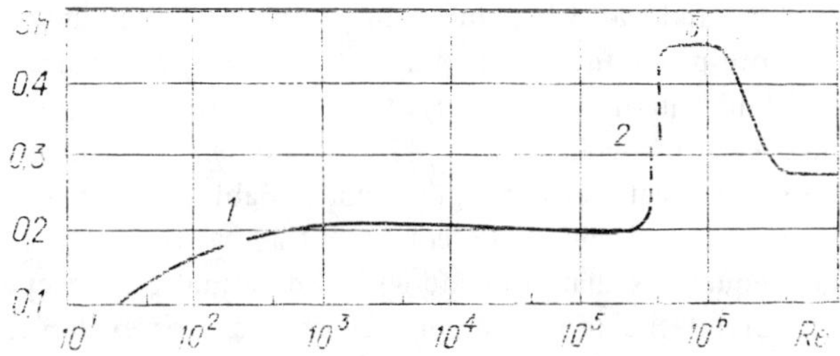

FIG. 4.11 Relation of the Strouhal number and the Reynolds number for a cylinder in crossflow: 1, the transitional region; 2, the separation bubble region; 3, the region of two separation bubbles.

The relationship between the Strouhal number and Reynolds number (Fig. 4.11), which results from numerous studies on a circular cylinder, shows a number of distinct regions. Vortex street formation in the wake starts as early as at $Re \approx 50$. The street of staggered vortices is 100 diameters long and is laminar in nature. In the range of Re from 150 to 300, flow disturbances and loss of regularity occur. With a further increase of Re (from $3 \cdot 10^2$ to $2 \cdot 10^5$), the regular vortex shedding is restored, and a constant value of $Sr = 0.2$ is established. However, the regular vortex street is now shorter and occupies the distance of only about 50 diameters from the cylinder and the wake becomes fully turbulent. On establishment of the critical flow regime, the drag coefficient decreases sharply, and the Strouhal number increases to the value of 0.46 (Fig. 4.11). The decrease in drag coefficient is related to the downstream shift of the separation point and to the narrowing of the wake. In critical flow on a cylinder, a two-step increase of the Strouhal number has been noted [93] and related to the separation bubble. The first increase (up to $Sh = 0.32$) was observed when a separation bubble was formed on one side of the cylinder. With the appearance of a separation bubble on the other side of the cylinder, the second increase to $Sh = 0.46$ was observed. At $Re > 6 \cdot 10^5$, regular vortex shedding is again destroyed, but it reappears once more at $Re = 3.5 \cdot 10^6$ [27], when $Sh = 0.27$. The drag coefficient is somewhat increased because the wake becomes wider with the upstream shift of the separation point.

We conclude that regular vortex shedding exists over a wide range of Re, and provides a constant exchange of mass, momentum, and energy between the vortex wake and the free stream. Mass transfer is dominated by vortex action over nearly the whole range, molecular exchange processes being important only at $Re < 40$.

In addition to their influence on transfer processes, wake phenomena have a direct influence on the vibration of the tubes involved. A single cylinder, or a bank of tubes, may be considered as an elastic body with a specific natural frequency. Interaction of the cylinder with the vortices, which are shed intermittently from its two sides, arises because of the asymmetric character of the flow giving rise to longitudinal and transverse fluctuating forces on the cylinder. The transverse forces are more pronounced, and when the frequency of vortex shedding coincides with cylinder natural frequency, the amplitude of vibration starts to increase and may even cause its failure. This phenomenon is an important one in the design of the heat transfer equipment. An ability to predict the Strouhal number for vortex shedding in the equipment allows the designer to avoid vortex shedding frequencies that coincide with the natural frequency. Values of Sh are known for all the most common geometries, for example, Sh = 0.2 for a circular cylinder. Recently, research has been carried out with the objective of obtaining more general design equations for whole classes of bodies. A relationship between the Strouhal number and the hydraulic drag has been proposed [94], derived from the dynamic behavior of the separation points and of the vortex street. For a wide range of bodies in crossflow, the relation was expressed in [94] as

$$c_w \,\mathrm{Sh} = 0.26\,(1 - e^{-2.38\,c_w}) \tag{4.5}$$

An approximate solution of Eq. (4.5) is presented in Fig. 4.12.

Bearman [93] and Roshko [95] related Sh to other parameters of the wake. From [95], $\mathrm{Sh}_R = fh/U$, where h is the distance between the separated vortex streets, and U is fluid velocity outside the boundary layer upon separation. Here $\mathrm{Sh}_R = 0.163$ applies for a large number of blunt bodies. In the other study [93], h was replaced by b, the distance between the external layers of the vortex streets at the point of their extinction, and resulted in $\mathrm{Sh}_B = fb/U = 0.27$. A number of other prediction techniques have been suggested, but in our opinion, Eqs. (4.4) and (4.5) are most convenient for practical calculations.

Fluid dynamics in the wake is affected strongly by the free-stream turbulence and by the blockage factor. The results of Akylbaev et al. [96] suggest the primary influence of these factors is on the length of the circulation zone. An increase of k_q and Tu leads to a shorter circulation zone and to an irregular vortex shedding. They also suggest [96] a relationship between Sh, blockage factor, and the frequency of vortex shedding:

$$\mathrm{Sh} = 0.2\,(1 + 7.25\,k_q^{3.3}) \tag{4.6}$$

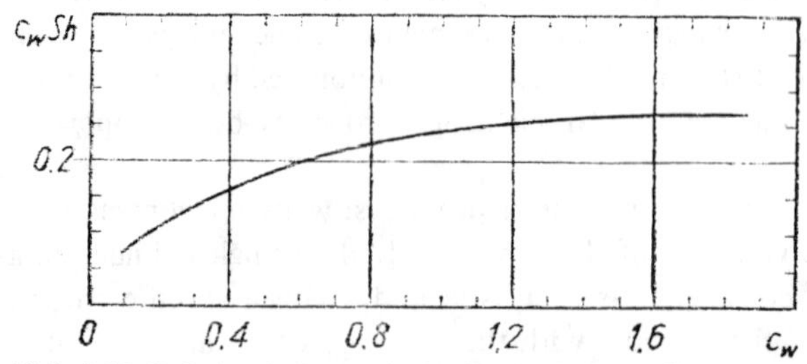

FIG. 4.12 Relation between the hydraulic drag coefficient and the Strouhal number.

The relationship applies to the subcritical flow regime ($2 \cdot 10^4 < \mathrm{Re} < 1.2 \cdot 10^5$), and it is implicit that Sh is independent of Re.

According to the correlation of Devnin [94], streamlining of bodies at subcritical Re results in a reduction in shedding frequency and hence Sh. For a cylinder of elliptic cross section with a 1:2 axial ratio, Sh = 0.13 and 0.19 for flows parallel to the major and to the minor axis, respectively. In the former case, Sh = 0.6 for the critical flow regime, which is much higher than for a circular cylinder. The change is probably due to the upstream shift of the separation point noted on bodies that produce wide wakes. In the subcritical flow regime, boundary-layer separation occurs at the rear stagnation point on an elliptic cylinder.

4.4 SHEAR STRESS DISTRIBUTION

The shear stress on the cylinder surface is an important fluid dynamic factor. The prediction of its distribution gives an insight into the development and dynamics of separation of the boundary layer.

With a cylinder in crossflow, the curvature of the surface gives rise to a longitudinal pressure gradient $dp/dx \neq 0$ in the boundary layer. It follows from the momentum equation,

$$u \frac{\partial u}{\partial x} + v \frac{\partial u}{\partial y} = -\frac{1}{\rho} \frac{dp}{dx} + \frac{1}{\rho} \frac{\partial \tau}{\partial y} \tag{3.1}$$

that, for the boundary conditions $y = 0$ and $u = v = 0$, the shear stress and pressure are related by $(\partial\tau/\partial y)_{y=0} = dp/dx$. This suggests that any change in the pressure gradient, be it in the front or in the rear, is reflected by a change in the shear-stress distribution. Thus in the front part, where $dp/dx < 0$, the shear stress decreases with the distance from the cylinder surface. In the rear, where $dp/dx > 0$, it increases up to a certain distance, then starts to decrease, and approaches zero in the outer boundary layer.

Shear stress at the wall is related to the velocity gradient,

$$\tau_w = \mu \left(\frac{\partial u}{\partial y} \right)_{y=0} \tag{4.7}$$

Thus any change of the shear stress τ_w gives a corresponding change in the velocity gradient $\partial u/\partial y$.

In a geometrical representation (Fig. 4.13*a*), the shear stress may be represented by the tangent of the angle formed by a normal to the cylinder surface and a tangential to the velocity profile. With $\tan \beta > 0$ (Fig. 4.13*b*), the velocity gradient is positive, and the shear stress is above zero. Boundary-layer separation is assumed to occur when $(\partial u/\partial y)_{y=0} = 0$ and $\tan \beta = 0$. Thus, at the separation point,

$$\tau_w = \mu \left(\frac{\partial u}{\partial y} \right)_{y=0} = 0 \tag{4.7a}$$

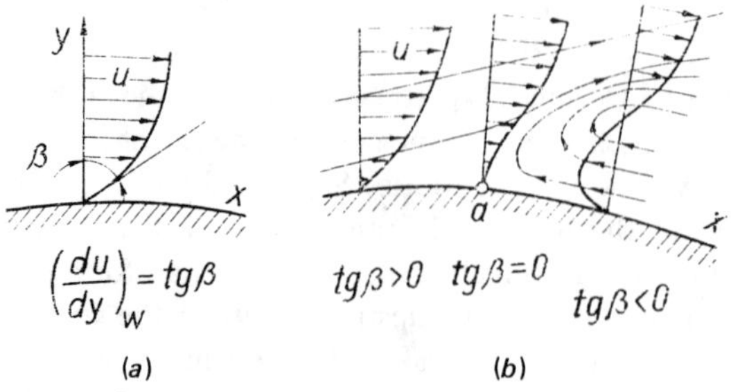

FIG. 4.13 Relationship between the distribution of the velocity and the shear stress: (*a*) determination of the shear stress, and (*b*) velocity distribution in the vicinity of the boundary-layer separation.

i.e., the shear stress—or in other words, the friction drag—is zero, and the fluid stream can leave the surface.

At a certain point downstream of the separation, $\tan\beta < 0$ (Fig. 4.13*b*), and flow reversal begins and the shear stress is negative. The velocity near the wall is negative relative to the main flow. This reverse stream is opposed by the boundary layer flow and begins to coil itself, giving rise to a vortex.

In spite of the rather complex nature of the flow pattern, shear stress was successfully modeled in the early studies by making a number of assumptions (Blasius [97], Howarth [98], and Hiemenz [82]). For the prediction of shear-stress distribution, of the velocity distribution in the boundary layer, and of the boundary-layer separation point, these investigators used analytical integration of the boundary-layer equations. The pressure distribution used was either calculated for inviscid, ideal flow (Fig. 4.1, curve 1) or was derived from measured values. More precise, numerical integration of the boundary-layer equations for a cylinder in crossflow has been recently carried out by Schönauer [99]. The experimental results and numerical integration results were close to the values obtained from the approximate equations. With the exception of reference [40], which refers to the subcritical flow, the early studies ignored such factors as the free-stream turbulence and the blockage factor. It was for this reason that studies of these factors have been included in the present study of the shear stress on elliptic and circular cylinders in flows of air, water, and transformer oil.

The Circular Cylinder

The experiments on circular cylinders were performed in collaboration with Ruseckas [100]. A cylinder of $d = 50$ mm was used in different fluids, keeping the blockage factor constant at the moderate value $k_q = 0.25$. No correction was made for the effects of the channel walls. The shear-stress distribution and the pressure distribution on the surface were analyzed simultaneously because of their very close interaction and because a general view of the processes was sought.

The shear stress on the wall, due to the friction drag, may be expressed in terms of the viscosity and the velocity gradient at the wall,

$$\tau_w = \mu \frac{\partial u}{\partial y} \tag{4.7}$$

It is convenient, in what follows, to express τ_w in the nondimensional form $\bar{\tau}$, defined as follows:

$$\bar{\tau} = \frac{2\tau_w}{\rho U_\infty} \sqrt{\mathrm{Re}} \tag{4.8}$$

As is seen from Figs. 4.14–4.17, $\bar{\tau} = 0$ at the front stagnation point. In the subcritical regime, $\bar{\tau}$ reaches a maximum at $\varphi \approx 60°$ and the location of the

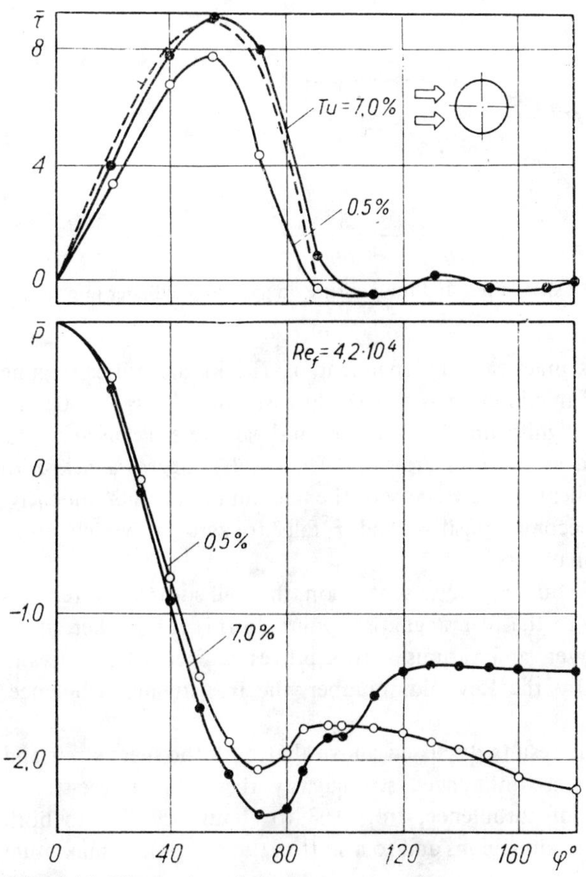

FIG. 4.14 Distribution of the shear stress and the pressure on a circular cylinder in air. (Dashed line shows analysis.)

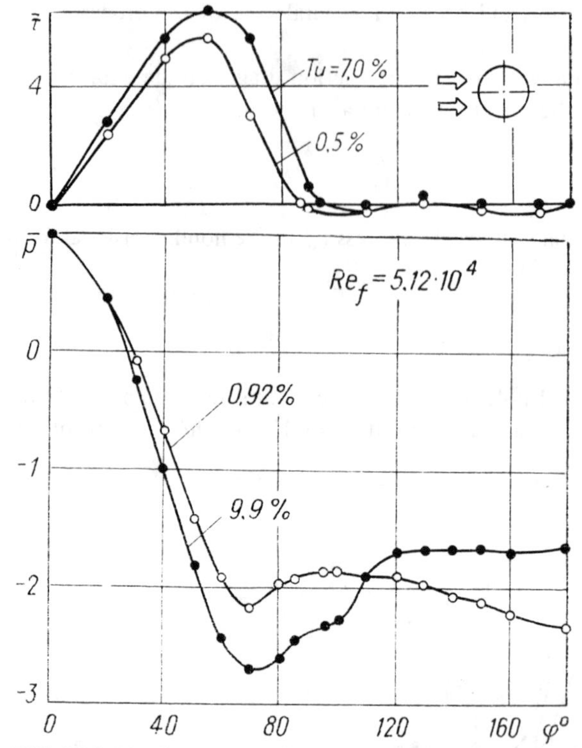

FIG. 4.15 Distribution of the shear stress and the pressure on a circular cylinder in water.

maximum shear stress is practically independent of Tu. In the critical regime (Fig. 4.17) the point of maximum $\bar{\tau}$ is shifted downstream to $\varphi = 70\text{–}80°$ and the effect of Tu is more significant. The increase in shear stress is caused by the acceleration of the fluid as the pressure falls. The results suggest a near-zero value of the pressure gradient in the vicinity of the maximum in $\bar{\tau}$. Subsequently, the pressure gradient becomes positive and $\bar{\tau}$ falls to zero, at which point boundary-layer separation occurs.

Beyond the point of boundary-layer separation, the wall shear stress remains close to zero ($\bar{\tau} = 0$) up to the rear stagnation point. In this region there is no steady-state boundary layer, and various vortex processes occur (periodic and nonperiodic), governed by the Reynolds number, the free-stream turbulence, and other factors.

It is evident from the results discussed above that both the shear stress and the pressure distribution are influenced strongly by the level of free-stream turbulence. An increase of turbulence, from 0.3 to about 10%, led to both higher shear stresses in certain regions and to a shift in the position of maximum shear stress.

Predictions using the analytical techniques described in Chap. 3 are also shown in Fig. 4.14. The predicted increase in shear stress is mainly caused by the deformation of the velocity profiles in the boundary layer, a higher velocity gradient at the wall and a thicker boundary layer (Fig. 3.5). In the region of decreasing shear stress preceding the separation point, the main factor is that of decreased pressure gradient.

The fact that the shear stress in the region near the rear of the cylinder is only slightly affected by the free-stream turbulence may be explained by the insensitivity of the highly turbulent flows in the wake to the minor external disturbances. At the front stagnation point the shear stress remains close to zero over the full range of free-stream turbulence studied. Judging by the location of the zero shear-stress point ($\bar{\tau} = 0$) in the subcritical flow regime, the separation point is shifted upstream with an increase of turbulence. In the near-critical regime (Fig. 4.16), the shift is more pronounced ($\varphi = 90$–$120°$).

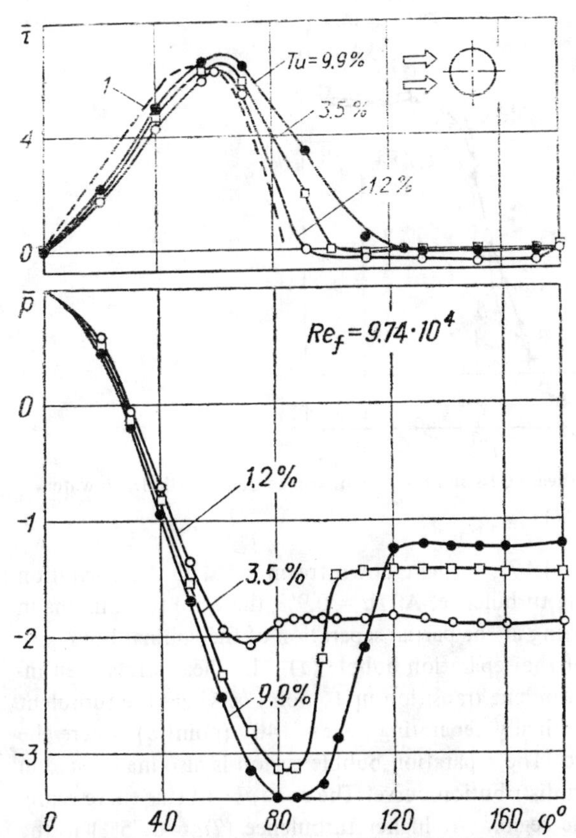

FIG. 4.16 Distribution of the shear stress and the pressure on a circular cylinder in water for different flow regimes, with dashed line from reference [26].

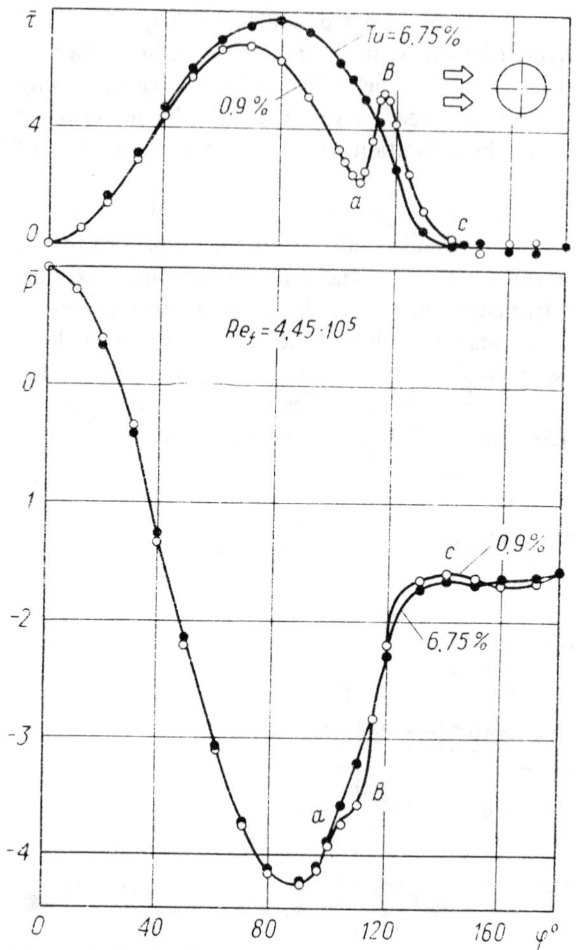

FIG. 4.17 Distribution of the shear stress and the pressure on a circular cylinder in water for the critical flow regime.

In the critical regime (Fig. 4.17), a slight downstream shift of the separation point occurs with increasing turbulence. At $Tu = 0.9\%$, there are two minima in shear stress, the first occurring at the partial separation of the laminar boundary layer and the formation of the separation bubble (*a*). The shear stress then increases due to a laminar-turbulent transition up to point *b*, where the turbulent boundary layer reattaches, finally separating at $\varphi \approx 140°$ (point *c*), where the shear stress approaches zero. The separation bubble region is also manifested in the zone *ab* in the pressure distribution curve. The zone *abc* of this curve represents the separation bubble region. At higher turbulence ($Tu = 6.75\%$) in the critical flow regime, the laminar-turbulent transition occurs in the boundary layer without the formation of the separation bubble (see Fig. 4.17).

The results shown in Fig. 4.16 agree well with those from the earlier study by Achenbach [26] in air. Minor discrepancies in the front stagnation region may be explained by the very different turbulence levels in the two studies ($Tu = 0.45\%$ in reference [26], compared with a minimum value of Tu of 1.2% in our studies).

In practical heat transfer operations, there exists a temperature difference between the surface and the fluid. In the present work, experimental investigations were made of temperature difference effects on shear stress in flow over heated cylinders in air, and analytical predictions were made for air and other fluids for both heating and cooling.

The experimental results for shear stress in a flow of air are shown in Fig. 4.18. The shear stress increased with increasing temperature differences, probably due to increasing viscosity of the air as the temperature increases.

The analytical predictions were made using the methods described in Chap. 3 and gave similar variations of the shear stress in air (Fig. 4.19, $\mathrm{Pr} = 0.71$). For liquids ($\mathrm{Pr}_f = 3.5$ and 242), shear stress decreased with increasing temperature difference due to the reduction of viscosity with temperature. The effect of the heat flux direction was also different for liquids and air, as shown in Fig. 4.19.

The Elliptic Cylinder

Shear stress distribution on an elliptic cylinder with 1:2 ratio between major and minor axes, and with air flow parallel to either axis was studied. Figures 4.20-

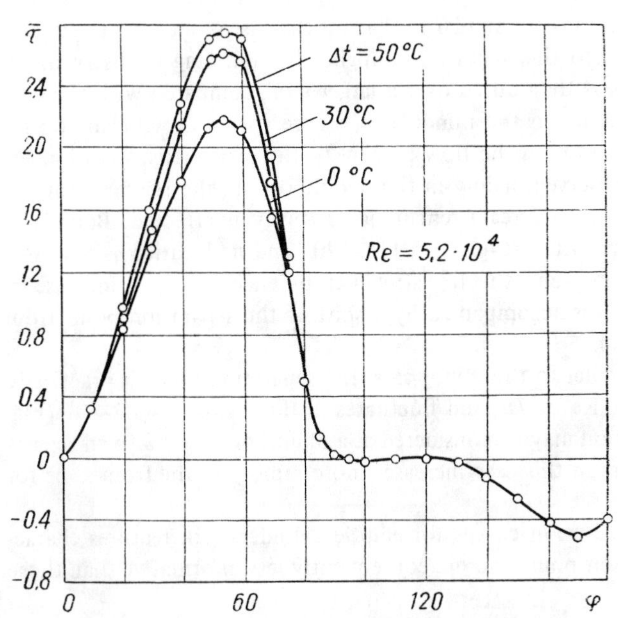

FIG. 4.18 The effect of the temperature difference on the shear stress in the flow of air (experiment).

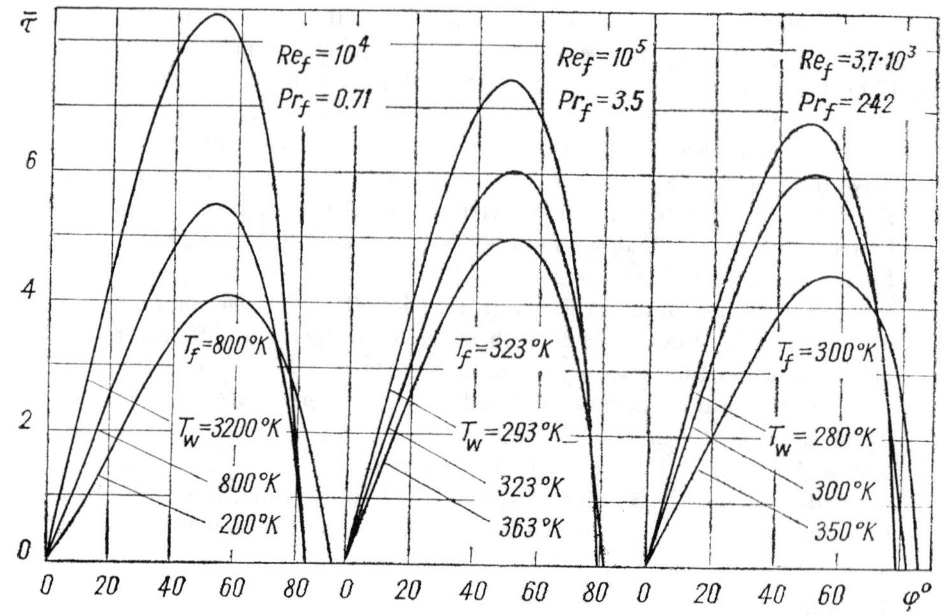

FIG. 4.19 Analytical predictions of the effect of the temperature difference on shear stress.

4.22 suggest that, as with the circular cylinder, an increase in shear stress and a decrease of pressure occurs with the increasing free-stream turbulence. The physical interpretation must be similar to that for the case of a circular cylinder. With the flow parallel to the major axis a higher friction drag is observed, because the shear stress is distributed over a larger area, compared with the case where the flow is parallel to the minor axis, where vortex flow occurs over a large part of the surface. With the flow parallel to the major axis, variations of the shear stress are observed mainly in the middle region, the regions near the stagnation points (front and rear) remaining relatively unaffected. Boundary-layer separation occurs in the region $\varphi = 110\text{–}140°$ and its location is governed by the Reynolds number and the free-stream turbulence. Thus, an increase of *Tu* from 0.76 to 9.9% is accompanied by a shift of the separation point from 100 to 140° (Fig. 4.21).

With the flow parallel to the minor axis, the separation point is practically independent of either Re or *Tu*, and fluctuates in the region $\varphi = 87\text{–}89°$ (Fig. 4.22). This configuration may be considered as a blunt body with a fixed separation point. Shear stress in this case increases more rapidly in the front zone for $\varphi > 40°$.

The pressure distribution curves for elliptic cylinders lack features characteristic of the separation point and are consequently less informative than those for circular cylinders.

The pressure distribution obtained for an elliptic cylinder with higher

Reynolds number flow parallel to the major axis suggests that the cylinder is behaving as a "slender body." Thus, there is a downstream shift of the boundary-layer separation point and higher pressures in the rear part (Figs. 4.20 and 4.21). For an elliptic cylinder with the flow parallel to the minor axis, the pressure continues to fall in the rear zone, this behavior being only slightly affected by Re.

4.5 PRESSURE DISTRIBUTION

The Circular Cylinder

The distribution of pressure on the cylinder surface described by Eq. (2.11) constitutes one of the important structural parameters of the flow. Its prediction opens the way for the determination of the velocity distribution in the front part outside the boundary layer, and of the friction drag, which is the

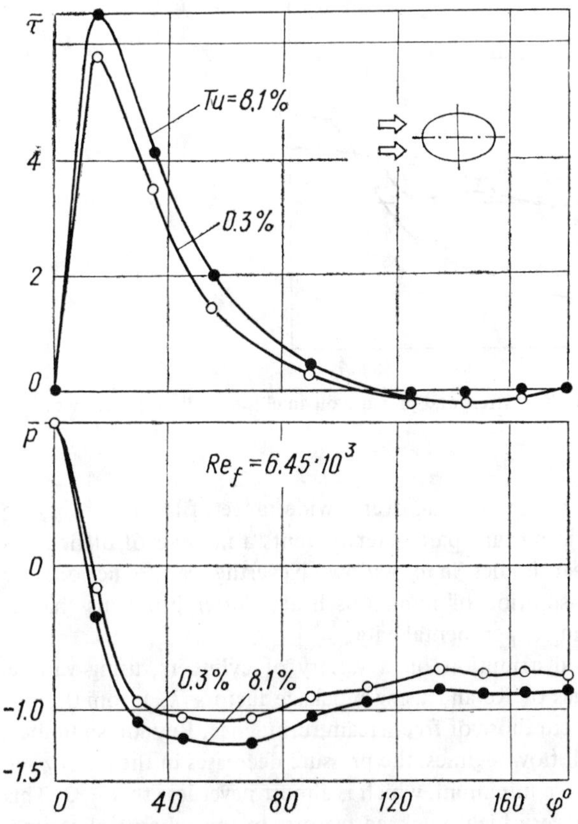

FIG. 4.20 Distribution of the shear stress and pressure on an elliptic cylinder in flow of transformer oil parallel to the major axis.

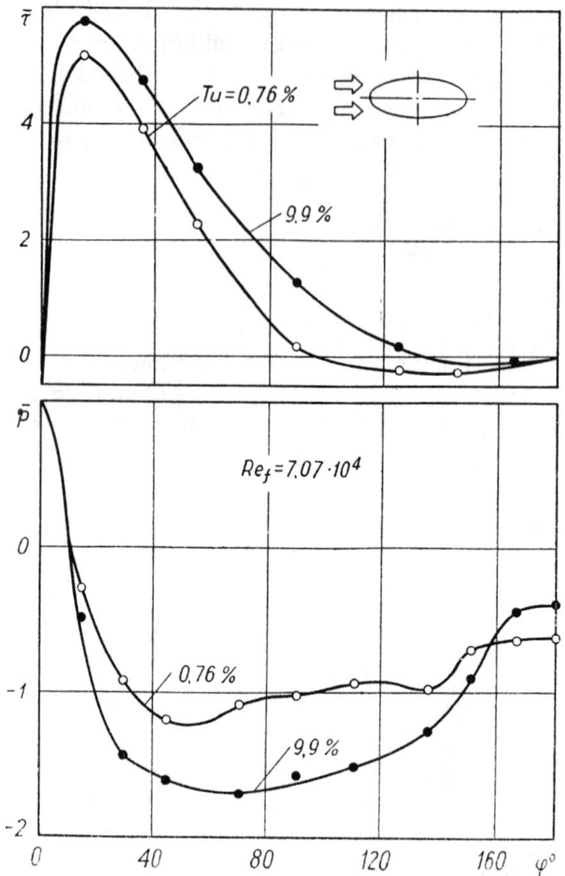

FIG. 4.21 Distribution of the shear stress and pressure on an elliptic cylinder in flow of air parallel to the major axis.

largest component of the hydraulic drag over a wide range of Re. However, the prediction of this highly important parameter presents a number of difficulties for the case of curvilinear bodies in crossflow. Nevertheless, predictions are urgently needed for the solution of numerous heat-transfer problems that at present must be solved using experimental data.

We studied pressure distribution on a variety of cylinders, using various fluids, covering wide ranges of Re and using blockage factors (k_q) from 0.25 to 0.7. The effects studied were those of free-stream turbulence, Reynolds number, and blockage factor. In all flow regimes, the pressure decreases in the front zone of the cylinder and falls to a minimum, which is almost never less than -3. This limit is only exceeded for very high blockage factors. In the subcritical regime, the minimum is a shallow one at $\varphi \approx 70°$ (Figs. 4.14, 4.16, 4.23). The minimum pressure is low in the subcritical regime because of the early boundary-layer separation and the wide wakes. For the lower range of Re, an increase of turbulence (Figs. 4.14, 4.15: at $Tu = 7$ and 9.9%) has little effect on the location of the minimum pressure, but the boundary layer becomes thicker, and the pressure decreases as far as about the separation point, and only then starts to increase again. This results in a shift of the separation point and in a higher base pressure. A simultaneous increase of Re and Tu provides a downstream shift of the minimum pressure to $\varphi \approx 85°$ (Figs. 4.16, 4.17, 4.23, 4.25, 4.26). In water, a shift of the minimum to $\varphi \approx 80°$ was observed with an increase of Tu (Fig. 4.26). An increase of the blockage factor causes the minimum pressure point to shift to $\varphi \approx 100°$ (Fig. 4.27).

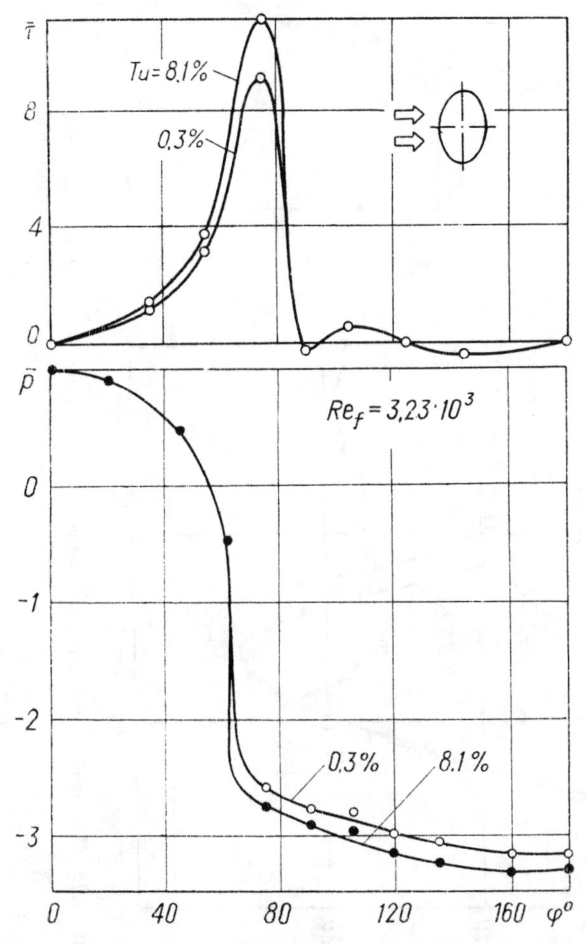

FIG. 4.22 Distribution of shear stress and pressure on an elliptic cylinder in flow of transformer oil parallel to the minor axis.

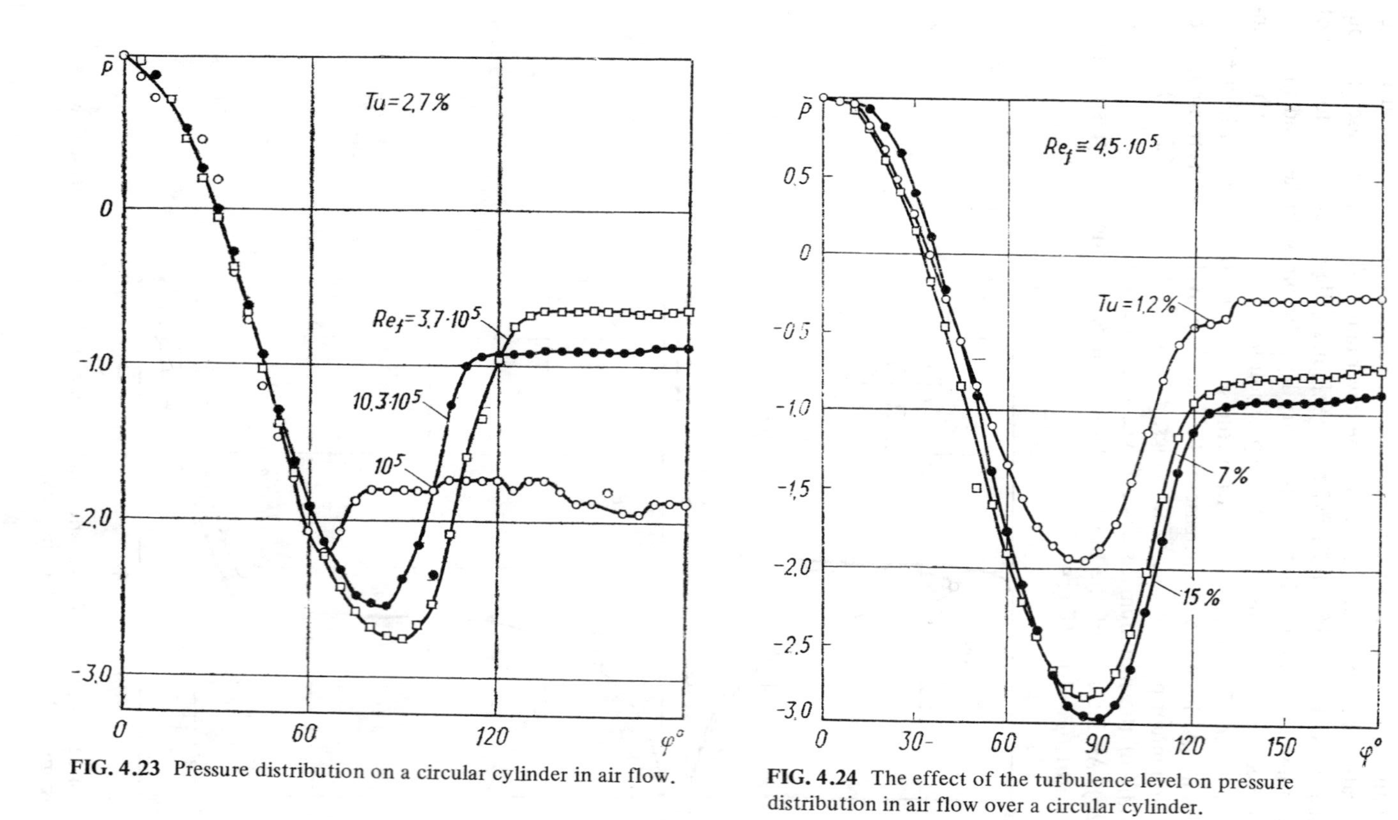

FIG. 4.23 Pressure distribution on a circular cylinder in air flow.

FIG. 4.24 The effect of the turbulence level on pressure distribution in air flow over a circular cylinder.

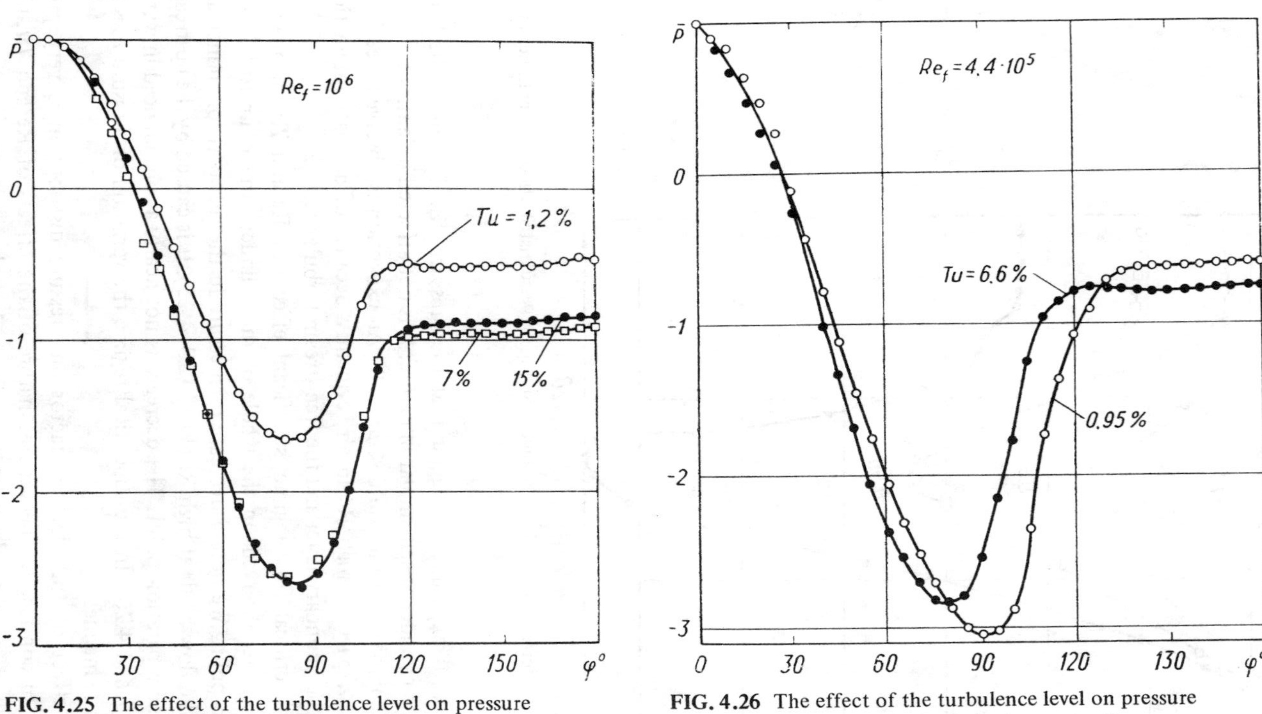

FIG. 4.25 The effect of the turbulence level on pressure distribution on a circular cylinder in air flow at high values of Re.

FIG. 4.26 The effect of the turbulence level on pressure distribution over a circular cylinder in a water flow.

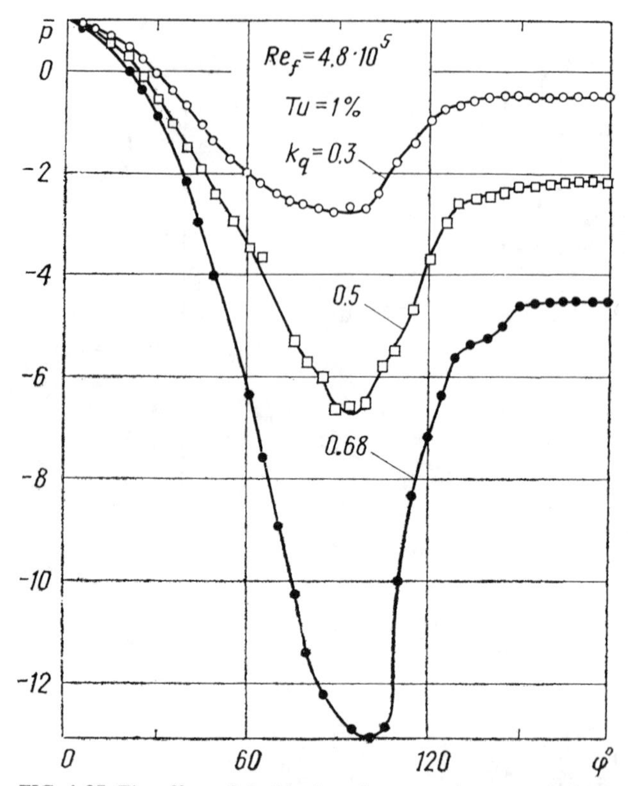

FIG. 4.27 The effect of the blockage factor on the pressure distribution on a cylinder in water flow.

Numerous experiments suggest that the pressure distribution in the front part of the cylinder is self-similar in nature. It is not affected by either Re or *Tu* over the approximate interval $0 < \varphi < 40°$. An exception is observed (see Figs. 4.24 and 4.25) with high Re in air, evidently because of a closer interaction between the boundary layer and the free-stream turbulence.

In the critical flow regime, significant effects of Re and *Tu* on pressure distribution are observed on the whole of the cylinder surface, up to the rear stagnation point. We note a decrease of pressure in the rear part for an increase of *Tu* in the flow of air at high Re (Figs. 4.23-4.25). It is caused by the upstream shift of the separation point. The decrease is not nearly as pronounced in water (Figs. 4.17 and 4.26). In the subcritical regime the variation of pressure goes in the opposite manner.

The effect of the blockage factor on pressure distribution is very pronounced. In contrast to the variable nature of the effects of Re and *Tu*, the effect of k_q is consistently evident on the pressure distribution over the whole surface (Fig. 4.27). The pressure decreases sharply with an increase of the blockage factor. With a change of k_q from 0.3 to 0.68, the pressure becomes over four times lower, and even nine times lower in the rear part. The influence of blockage is exerted through the increased velocity gradient on the front part, as seen in both experimental and analytical curves of the velocity distribution (Fig. 3.2). The general tendencies of the effects of Re and *Tu* were not changed with the increase of k_q.

The Elliptic Cylinder

The reader can understand the significant features of the pressure distribution on the elliptic cylinder from the description (Section 4.4) of the shear-stress distribution. We only note here that, at high Re, the dynamic behavior of pressure in the rear part is related more to the vortex flow in the near wake (Figs. 4.20 and 4.22) than to the dynamics of the boundary-layer separation or of the width of the wake. Since the separation point is stable on the elliptic cylinder, it has no influence in changing pressure distribution as *Tu* increases.

4.6 VELOCITY DISTRIBUTION IN THE BOUNDARY LAYER

The development of the velocity distribution deserves careful attention because of its importance in calculations. For the boundary layer on a cylinder in crossflow, the Bernoulli equation is

$$p_\infty + \frac{\rho U_\infty^2}{2} = p_\varphi + \frac{\rho U_\varphi^2}{2} = \text{const} \tag{4.9}$$

and the velocity in the boundary layer is given by

$$U_\varphi = U_\infty \sqrt{1 + \frac{2\,(p_\infty - p_\varphi)}{\rho U_\infty^2}} \tag{2.12}$$

Using the measured values of the pressure drop (the difference in pressure between the free stream and the surface) and of the free-stream velocity, the fluid velocity outside the boundary layer can be determined from Eq. (2.12) as a function of the angular distance up to the separation point.

For the velocity outside the boundary layer in an infinite flow, the following equation was suggested by Frössling [18] and Hiemenz [82]:

$$\frac{U_\varphi}{U_\infty} = 3.6314\left(\frac{x}{d}\right) - 2.1709\left(\frac{x}{d}\right)^3 - 1.514\left(\frac{x}{d}\right)^5 \tag{4.10}$$

where x is the arc length from the front stagnation point.

In practical applications, fluid flows over cylinders are influenced by the channel walls. Our experiments and analytical solutions with different blockage factors resulted in the following equation for the velocity outside the boundary layer:

$$\frac{U_\varphi}{U_\infty} = A_1^* \frac{x}{d} + A_2^* \left(\frac{x}{d}\right)^3 + A_2^* \left(\frac{x}{d}\right)^5 \tag{4.11}$$

where

$$A_1^* = 3.631 \left(1 + \frac{k_q}{2}\right)$$

$$A_2^* = -2.1709\,(1 - 12.1284\, k_q^{2.226} + 3.737\, k_q)$$

$$A_3^* = -1.5144\,(1 - 18.542\, k_q^{2.277} - 6.878)$$

In the limiting case where $k_q = 0$, Eq. (4.11) is identical to Eq. (4.10).

Figure 4.28 presents the velocity profiles in air at different Re and *Tu* as derived from the predicted pressure distributions and from thermoanemometer measurements, respectively. The fluid velocity in the external part of the boundary layer increases with increasing Re and *Tu*. We suggest that this effect is due to deformation of the boundary layer or of the velocity profile within it.

Considerable insight can also be gained from the velocity distribution and the velocity profiles within the boundary layer. They may serve as indications of the boundary-layer development and separation. Any deformation of the velocity

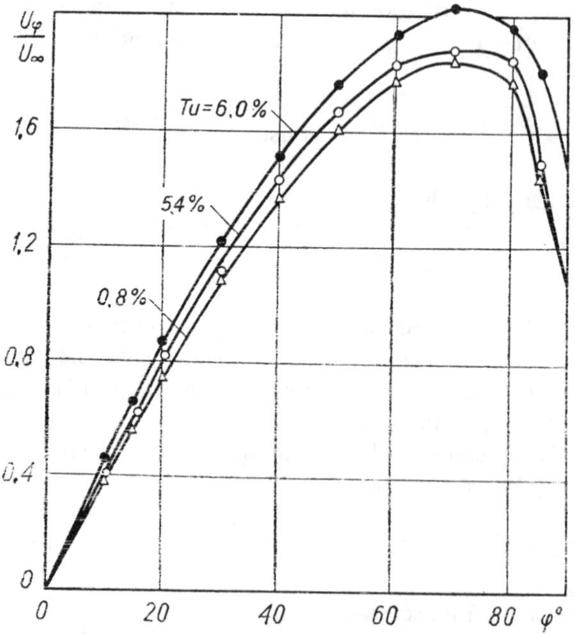

FIG. 4.28 Velocity distribution outside the boundary layer in air flow over a circular cylinder: $Tu = 0.54$ and 6.0% with $\mathrm{Re} = 6.7 \cdot 10^4$; $Tu = 0.8\%$ with $\mathrm{Re} = 3.5 \cdot 10^4$.

profiles is a reflection of a change in the boundary-layer thickness, be it due to the effect of the angular distance or to the effect of the free-stream turbulence. The measurement of the velocity profiles around the surface of a cylinder of small diameter presents numerous technical difficulties. Perhaps this fact has been the incentive for the development of the analytical technique, described in Chap. 3, for the prediction of the velocity distribution in the boundary layer and the effect of free-stream turbulence on it. The velocity profiles shown in Fig. 3.5, determined in cooperation with Vaitiekunas [101], indicate the significant deformation caused by the free-stream turbulence both in the external and in the near-wall parts of the boundary layer.

4.7 HYDRAULIC DRAG

The hydraulic drag on a cylinder is equal to the resultant force of friction P_f and pressure P_d acting on its surface [62]. The highly complicated dynamics of fluid flow around cylinders and its dependence on the Reynolds number have already been noted. The Reynolds number has also a large influence on the pressure drag. At a very low Re, in potential flow, the pressure force P_d is insignificant, and the pressure drag is governed by the friction force P_f.

From [83], at $\mathrm{Re} < 30$ the friction drag coefficient of a circular cylinder is

$$c_f = \frac{8\pi}{\mathrm{Re}\,(2.002 - \ln \mathrm{Re})} \tag{4.12}$$

With increasing Re, the inertia force begins to play an increasingly important role, until its contribution becomes equal to that of friction at $\mathrm{Re} \approx 30$. For Re in the range 30 to 10^4 [83], the friction drag is described by

$$c_f = 4\ \mathrm{Re}^{-0.5} \tag{4.13}$$

Over this range, the contribution of friction falls from 50 to 2% of the pressure drop. In our experiments with circular and elliptic cylinders in different fluids [100], the contribution of the friction drag to pressure drop was found to be in the range 3 to 1% for Re from $5 \cdot 10^3$ to 10^6, the contribution of the friction drag decreasing with Re. Similar results were obtained by Achenbach [26] for air flows.

Out studies of flow over an elliptic cylinder, with the flow parallel to the major axis, indicated that the contribution of the friction drag varied from 7 to 2% in the Re range from $2 \cdot 10^3$ to 10^5. With the flow parallel to the minor axis, the contribution was similar to that on a circular cylinder.

As will be seen from the above results, the friction effects are not the dominant ones in determining pressure losses in flow over cylinders. This contrasts with the case of turbulent flow over a flat plate, where the friction coefficient is widely applied in the analytical solutions and experimental interpretations on the basis of the Reynolds analogy. The direct determination of the friction component of the hydraulic drag is only feasible at $Re < 10^3$. At $Re > 10^3$, the frictional drag and the location of the separation points are determined indirectly on the basis of shear-stress distribution measurements.

The pressure drag component of the hydraulic drag in flow over a cylinder is governed by inertial effects in the medium and high ranges of Reynolds number where it dominates the hydraulic drag. The variation with Re of the hydraulic drag coefficient on circular and elliptic cylinders is illustrated in Fig. 4.29 for infinite flows with low turbulence. The curves are derived from a wide range of data obtained by different authors. In the subcritical regime ($10^4 < Re < 2 \cdot 10^5$), the hydraulic drag coefficient of the circular cylinder is $c_D = 1.2$ and that of the elliptic cylinder is $c_D = 0.7$ and 1.7, for the flows parallel to the major and the minor axes, respectively. In the critical flow regime, the pressure drag of the circular cylinder decreases to 0.3, increasing to 0.9 as the supercritical regime is entered. On the elliptic cylinder, for the flow parallel to the major axis, c_D is decreased to 0.1 in the critical flow regime.

In a separate study [102], pressure drag was measured under various conditions. The friction drag was ignored since its effect was insignificant for the fluid flow ranges and blockage factors studied.

The pressure drag coefficients were determined by integration of the static pressure distribution on the surface. Figure 4.30 presents the curves for the pressure drag coefficient in the subcritical regime at various turbulence levels. With an increase of turbulence, the pressure drag decreases as a consequence of the change in the pressure distribution, the influence of the near wake on the separation point, and the earlier laminar-turbulent transition. In turbulent flows, the critical regime is established earlier.

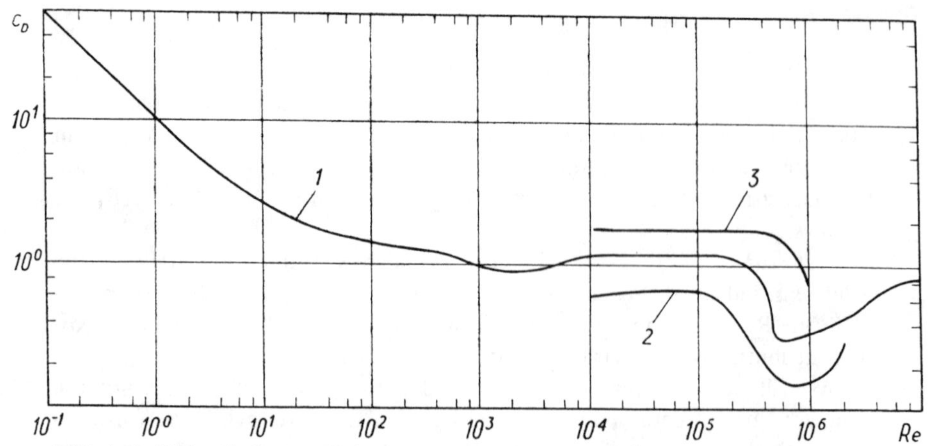

FIG. 4.29 Hydraulic drag coefficient as function of Re: 1, circular cylinder; 2, elliptic cylinder, flow parallel to the major axis; 3, elliptic cylinder, flow parallel to the minor axis.

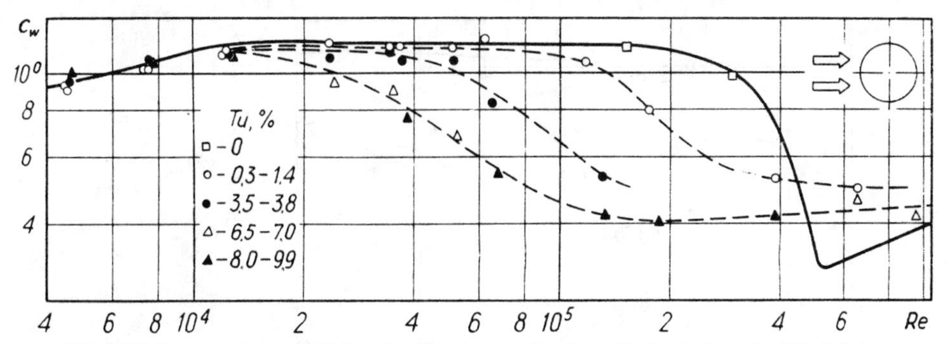

FIG. 4.30 Pressure drag coefficient for flow over a circular cylinder in the subcritical flow at various levels of turbulence.

We have attempted to interpret the results for the critical flow regime in terms of a modified Reynolds number $\mathrm{Re}_t = \mathrm{Re}\, Tu$ (see Fig. 4.31). The level of turbulence was varied from 0.3 to 10%. The data were fitted by the equation:

$$c_w = 1.09 \tag{4.14}$$

for $40 < \mathrm{Re}_t < 10^3$, and by the equation

$$c_w = 45.3\mathrm{Re}_t^{-0.51} \tag{4.15}$$

for $10^3 < \mathrm{Re}_t < 10^4$.

Figures 4.32 and 4.33 present the pressure drag coefficients for flow over the elliptic cylinder, for flows parallel to the major and the minor axis, respectively. With the flow parallel to the minor axis, the pressure drag is hardly affected by the free-stream turbulence, the larger part of the surface being under a vortex flow regime. The results agree closely with the findings of Devnin [94] (Fig. 4.32 and solid line in Fig. 4.33).

Figure 4.34 shows the effect of turbulence on the pressure drag in the critical flow regime. Here, with an increase of Tu, the critical flow regime is established at lower Re. The general effect of turbulence on the general level of the pressure drag is not large, the same trend being observed for both water and air flows.

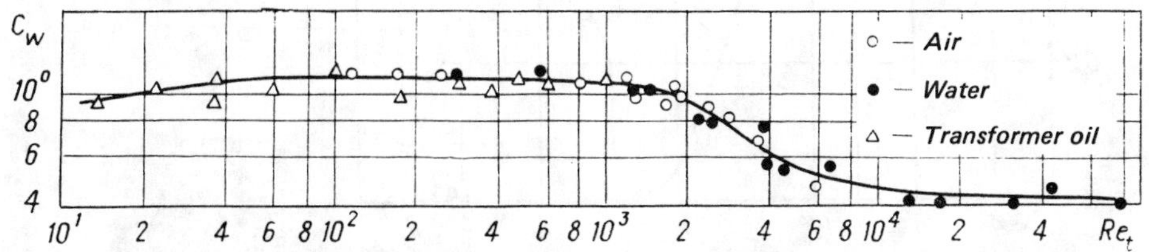

FIG. 4.31 Pressure drag coefficient for flow over a circular cylinder as a function of $\mathrm{Re}_t = \mathrm{Re}\, Tu$.

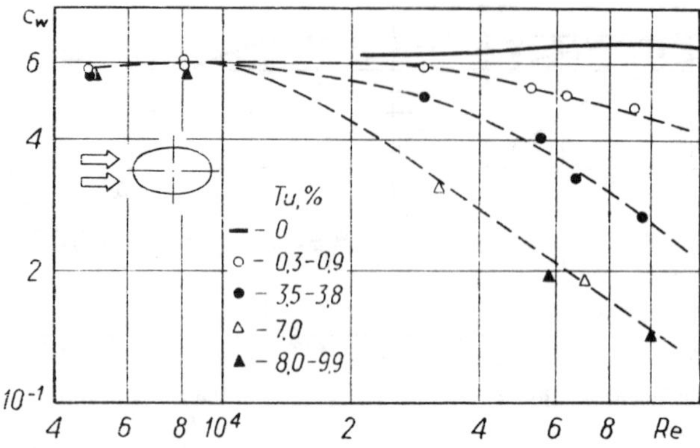

FIG. 4.32 Pressure drag coefficient for an elliptic cylinder with flow parallel to the major axis at various levels of turbulence.

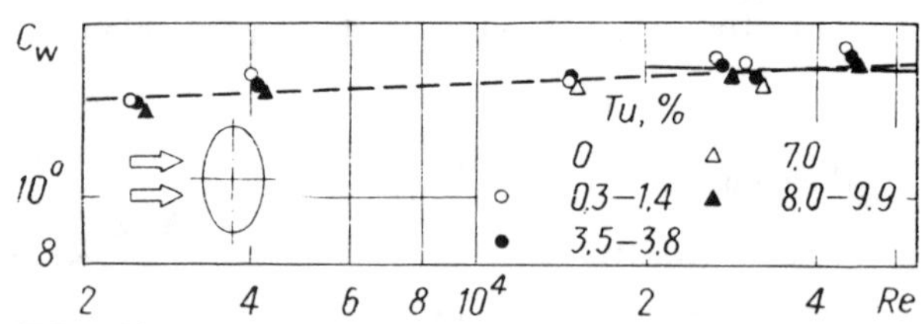

FIG. 4.33 Pressure drag coefficient for an elliptic cylinder with flow parallel to the minor axis at various levels of turbulence.

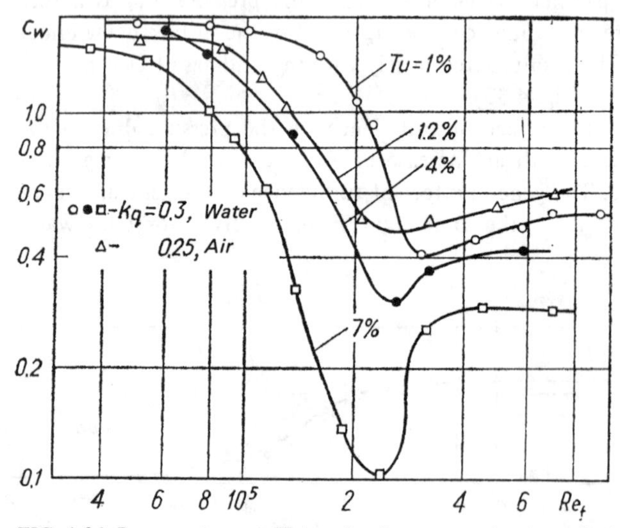

FIG. 4.34 Pressure drag coefficient for flow over a circular cylinder in the critical flow regime at various levels of turbulence.

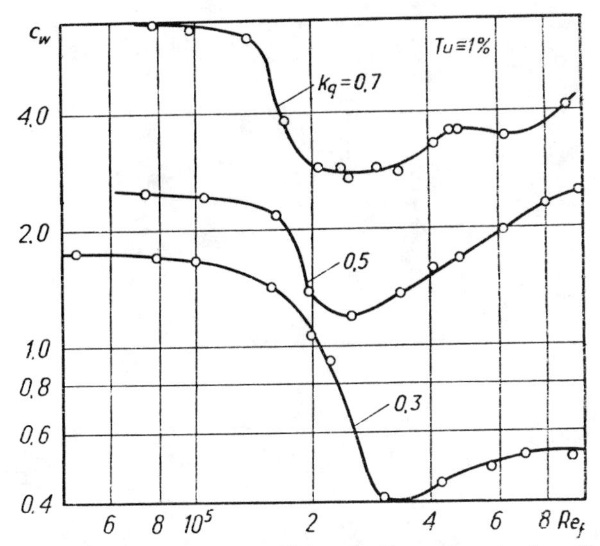

FIG. 4.35 Pressure drag coefficient for flow over a circular cylinder in the critical flow regime as a function of Re_f and blockage factor.

It will be noted that the pressure drag coefficients subcritical flow shown in Fig. 4.34 (left-hand side) are higher than those shown in Fig. 4.29. This is due to the higher blockage factor, Fig. 4.29 being for an infinite flow. A more systematic set of data for the effect of blockage factor is shown in Fig. 4.35. The pressure drag coefficient for the cylinder case in general increases with the blockage factor k_q. However, there is little effect of the blockage factor on the onset of the critical regime. For the range of Re_f covered in our work, the following empirical equation is suggested for the pressure drag c_w:

$$c_w = A_0 + A_1\left(0.5 - \frac{\operatorname{arc\,tg}\,[B_1\,(Re_f \cdot 10^{-5} - C_1)]}{\pi}\right) + A_2\left(0.5 + \frac{\operatorname{arc\,tg}\,[B_2\,(Re_f \cdot 10^{-5} - C_2)]}{\pi}\right) \tag{4.16}$$

This equation applies to both critical and supercritical flows but ignores the effects of turbulence. The values of constants are dependent on blockage factor k_q and are presented in Table 4.1.

TABLE 4.1 Values of Constants in Eq. (4.16)

k_q	A_0	A_1	B_1	C_1	A_2	B_2	C_2
0	0.1	1.11	1.8	3.95	0.25	0.80	6.0
0.3	0.22	1.58	2.5	2.06	0.30	1.10	4.5
0.5	0.88	1.62	5.8	1.78	1.50	1.20	4.3
0.7	2.25	4.80	6.0	1.58	1.90	1.22	4.2

CHAPTER

FIVE

HEAT TRANSFER AT THE FRONT STAGNATION POINT ON A CYLINDER

The fluid dynamic and heat transfer behavior in the vicinity of the front stagnation point on a cylinder may be considered as a special case of laminar or turbulent flow impinging normally onto a flat plate. This situation is observed in numerous practical situations, and the study of the fluid dynamic and heat transfer characteristics is highly important from both the practical and theoretical points of view. Normally impinging flows of fluids are observed on the front part of aircraft and rockets, in the thermal treatment of sheet materials produced in rolling mills, in some processes for drying, etc. The normally impacting flows may be deliberately introduced for the purpose of heat transfer augmentation or for thermal protection.

All the processes in the vicinity of the front stagnation point are conveniently simulated by studying flows on circular and elliptic cylinders, and the results can be extended to other situations. In the present chapter, the structure of the fluid dynamics in the vicinity of the front stagnation point is considered for various fluids, in order to evaluate the effects of fluid physical properties on the heat transfer. The other determining factors are the channel wall effect (expressed by the blockage factor), the integral longitudinal scale of turbulence, the Prandtl number, and the flow regime.

5.1 SPECIFIC FEATURES OF FLUID DYNAMICS AND HEAT TRANSFER AT THE FRONT STAGNATION POINT

The analytical technique described in Chap. 3, which takes account of the free-stream turbulence, yields parameters of the fluid dynamics and the heat transfer over nearly the whole front region. However, results for the front stagnation point $\varphi < \pm 1°$ cannot be obtained from this analysis. We consider here the analytical techniques for heat transfer and fluid dynamics at the front stagnation point (Fig. 5.1) and begin with the most general approach for normal impact on a flat plate. The velocity outside the boundary layer is expressed for the after-impact flow by a linear function

$$U_x = c_1 x \tag{5.1}$$

where c_1 is a constant.

The continuous increase of velocity causes a decrease of pressure. The flow and the heat transfer near the stagnation point are described by the boundary layer and energy equations (3.1-3.3) with the boundary conditions of Eq. (3.4). The free-stream velocity outside of the boundary layer is given by

$$u_1 = c_1 x \tag{5.2}$$

The potential flow outside the boundary layer is described by the Euler equation,

$$U_x \frac{\partial U_x}{\partial x} = -\frac{1}{\rho} \frac{\partial p}{\partial x} \tag{5.3}$$

which after the substitution of Eq. (5.2) becomes

$$U_x \frac{\partial U_x}{\partial x} = -\frac{1}{\rho} \frac{\partial p}{\partial x} = c_1^2 x \tag{5.4}$$

The momentum transfer equation, Eq. (3.1), can be reduced to

$$u \frac{\partial u}{\partial x} + v \frac{\partial u}{\partial y} = c_1^2 x + \nu \frac{\partial^2 u}{\partial y^2} \tag{5.5}$$

The velocity distribution to be described by Eq. (5.5) has to obey the following relationships:

$$u = x f'(y); \quad v = -f(y) \tag{5.6}$$

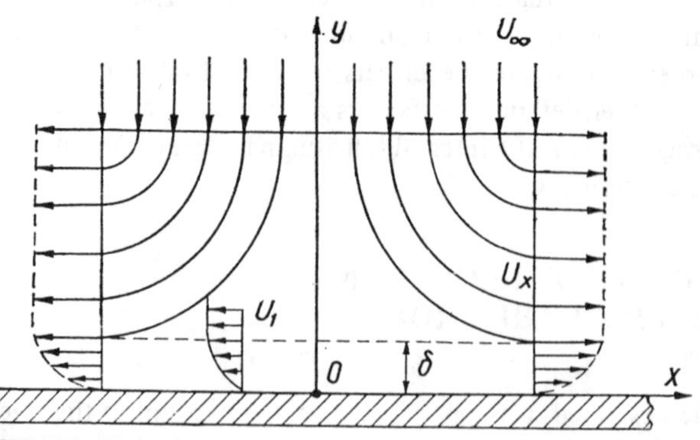

FIG. 5.1 Fluid dynamics in the front stagnation point on a screen.

To solve Eq. (5.5), we introduce the following auxiliary variables:

$$\eta = \sqrt{\frac{c_1}{\nu}}\, y \tag{5.7}$$

$$f(y) = \sqrt{c_1 \nu}\, \Phi(\eta) \tag{5.8}$$

The velocity distribution is then given by

$$u = c_1 x \Phi'(\eta), \quad v = -\sqrt{c_1 \nu}\, \Phi(\eta) \tag{5.9}$$

After the substitution of Eq. (5.9) into Eq. (5.5), and rearranging, we obtain the nonlinear differential equation

$$\Phi''' + \Phi\Phi'' - \Phi'^2 + 1 = 0 \tag{5.10}$$

where $\Phi = 0$ and $\Phi' = 0$ for $\eta = 0$, and $\Phi = 1$ for $\eta = \infty$.

This equation was solved by Hiemenz [82], who defined the Φ, Φ', and Φ'' terms as functions of η [53]. The Φ term represents the velocity distribution in the boundary layer at the front stagnation point:

$$\frac{U}{U_x} = \Phi'(\eta) \tag{5.11}$$

The boundary-layer thickness δ may be defined as that distance from the wall at which $\Phi' = U/U_x = 0.99$. From Eqs. (5.11) and (5.7) we have:

$$\delta = 2{,}4 \frac{\sqrt{\nu}}{c_1} \tag{5.12}$$

Thus, in the vicinity of the stagnation point, δ does not depend on the longitudinal coordinate x and is constant, its value depending on the fluid physical properties (i.e., on the kinematic viscosity).

To solve the energy equation, Eq. (3.2), an equation for the temperature distribution in the thermal boundary layer must be chosen. The temperature distributions may be described by the dimensionless variable

$$\Theta(\eta) = \frac{T - T_w}{T_f - T_w} \tag{5.13}$$

Substituting Eqs. (5.13) and (5.4) into the energy equation, and rearranging, we again arrive at a differential equation:

$$\Theta'' + \mathrm{Pr}\, \Phi\Theta' = 0 \tag{5.14}$$

where $\Theta = 0$ for $\eta = 0$, and $\Theta = 1$ for $\eta = \infty$.

Numerous solutions of this equation can be found in the literature. They all result in the following expression for heat transfer in the front stagnation point:

$$\mathrm{Nu}_x = 0{,}57\, \mathrm{Re}_x^{0.5}\, \mathrm{Pr}^{0.33} \tag{5.15}$$

$$\mathrm{Nu}_x = a(\mathrm{Pr}) \sqrt{\frac{U_x x}{\nu}} \tag{5.16}$$

where $a(\mathrm{Pr}) = 0.57$ for $\mathrm{Pr} = 1$ [53].

Alternatively, for the cylinder case, the velocity outside the boundary layer U_x may be replaced in Eq. (5.16) by the free-stream velocity U_∞, and the values of Nu and Re may be referred to the cylinder diameter d; in this case, we have the following expression for heat transfer in the stagnation region in flow over the cylinder:

$$\mathrm{Nu}_d = 1{,}14\, \mathrm{Re}_d^{0.5} \cdot \mathrm{Pr}^{0.33} \tag{5.17}$$

The technique suggested by Eckert [19] for prediction of the heat transfer on wedges is also applicable for cylinders. For a wedge, the velocity outside the boundary layer is expressed by the exponential function

$$U_x = c_1 x^{m_1} \tag{5.18}$$

With $m_1 = 1$, Eq. (5.18) coincides with Eq. (5.1) and represents the velocity distribution near the stagnation point. Equation (5.18) also describes the velocity distribution in the vicinity of the boundary-layer separation if a value $m_1 = -0.0804$ is selected.

To obtain a solution of a heat-exchange problem using Eq. (5.18), we employ the same boundary-layer equations (3.1-3.3) and the same boundary conditions (3.4), but different and more complex auxiliary variables [compared to those given by Eqs. (5.7) and (5.8)] are required. The solution obtained by Eckert for heat transfer between a fluid and a wedge was

$$\mathrm{Nu}_x = 0.56\, A\, \mathrm{Re}_x^{0.5} / \sqrt{2-\beta} \tag{5.19}$$

for Pr from 0.7 to 10. The effect of Pr was expressed by the exponential function

$$A = (\beta + 0.2)^{0.11}\, \mathrm{Pr}^{0.333 + 0.067\beta - 0.026\beta^2} \tag{5.20}$$

where $\beta = 2m_1/(m_1 + 1)$.

With $m_1 = 1$, Eq. (5.19) gives the heat transfer at the front stagnation point on a cylinder.

The above expressions cover only a limited range of Pr. To extend the range, we performed numerical analyses for Pr from 10 to 100 by an integral technique similar to that used by Eckert [19]. The resultant equation

$$\mathrm{Nu}_x = 0{,}570\, \mathrm{Re}_x^{0.5}\, \mathrm{Pr}^{0.35} \tag{5.21}$$

differs from that obtained by Eckert in the exponent of Prandtl number (0.35, not 0.33).

All the above equations [Eqs. (3.15), (5.19), and (5.21)] apply to the constant wall temperature boundary condition (t_w = constant). All cases of the heat transfer are dependent on the boundary conditions used. However, the effect of boundary conditions on heat transfer at the front stagnation point is, as yet, far from clear and there are few appropriate studies. Krall and Eckert [20] obtained numerical solutions of the problem by solving the boundary-layer equations for $\mathrm{Re} \leqslant 200$; they found that the heat transfer coefficient was not dependent on surface boundary conditions in the vicinity of the front stagnation point, up to $\varphi \approx 30°$. Further around the cylinder, higher heat transfer coefficients were observed with constant q_w than with constant t_w. We may reasonably expect that the region of independence of heat transfer coefficient on boundary conditions will be narrower at higher Re.

Our analytical studies were limited to the case of a laminar boundary layer on the front part of the cylinder. This situation is far from common in practical applications, where a certain degree of turbulence is usually observed in the free stream. Unfortunately, none of the available physical models can account analytically for the effect of free-stream turbulence on the heat transfer and fluid dynamics. However, models are available in which the boundary-layer equations are used together with idealized assumptions and empirical constants. One such model is that due to Kayalar [105] and illustrated in Fig. 5.2. It consists of a cellular structure with a defined wavelength λ_1 and a definite frequency of fluctuation, superimposed on a laminar boundary layer. Two vortices of opposite direction are assumed. The model was used for the prediction of heat transfer and the shear stress in the front stagnation region and incorporated a number of empirical constants.

A number of authors [106-108] relate the effect of turbulence to existence of vortices of the Taylor-Görtler type, which are observed near the front stagnation point and are caused by the reverse flow. Originally found on concave surfaces, these vortices were recently observed on convex bodies, including cylinders [53]. Kestin and Wood [106] suggest that, if these vortices exceed a critical size, they begin to interact with the boundary layer. Thus, when a

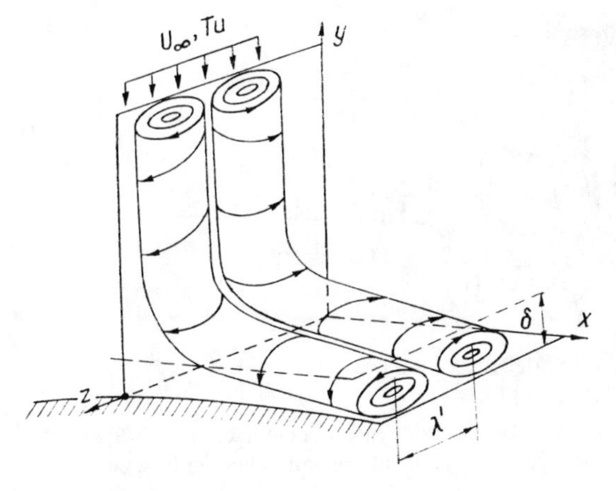

FIG. 5.2 Vortex cellular structure in the vicinity of the front stagnation point on a cylinder; model of Kayalar [105].

fluid impinges on a convex surface, inverse flows and vortices occur and interact with the boundary layer, resulting in a more intensive transfer. Based on visual observations, the following relation was obtained for the wavelength of the cellular vortex structure:

$$\frac{\lambda_1}{d} = \lambda_0 \pi \mathrm{Re}^{-0.5} \tag{5.22}$$

where $\lambda_0 = 1.56$, the wavelength of the cellular structure at $Tu \to 0$. The equation suggests a decrease of the wavelength with an increase of Re and Tu. The cellular vortex structure on circular cylinders in turbulent flows has been observed by visual, thermoanemometric [108], ablative, and other techniques. Figure 5.3 from reference [105] also supports the relationships discussed above between flow structure and Tu and Re.

Figure 5.4 illustrates the results of our observations of the fluid dynamic structure on a 250-mm-diameter cylinder placed in a water-filled open channel. Dyed fluid was injected along the front stagnation line in a water flow with low turbulence level and with $U_\infty = 0.04$ m/s. In confirmation of the results of other workers, we observed the cellular structure in the front region up to $\varphi = 60$-$70°$, that is, up to the separation point. The wavelength varied between 2 and 8 mm for the range of flow parameters studied, and corresponded to vortex diameters of 1 to 4 mm.

Previously, opinions on the location of the cellular structure have been contradictory: it is in the vicinity of the front stagnation point according to some authors, in the separation region according to others. We maintain that the boundary layer is penetrated by vortices over the whole region from the front stagnation point to the separation point. This conclusion is reached on the

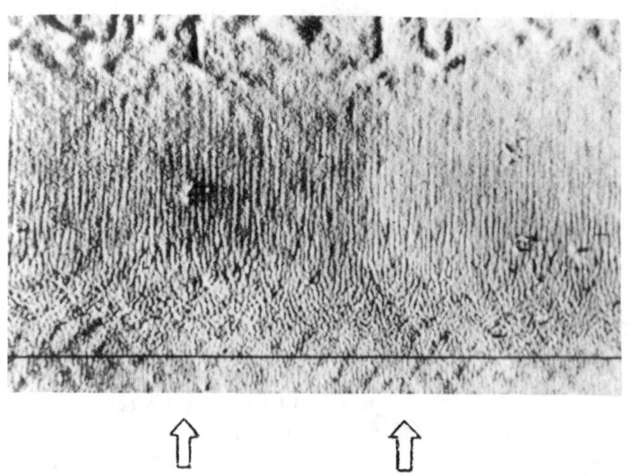

FIG. 5.3 Cellular fluid dynamic structure on the front part of a cylinder at $\mathrm{Re} = 9.4 \cdot 10^4$ and $Tu = 4.2\%$ [105]. The solid line represents the front stagnation line on the cylinder.

FIG. 5.4 Dynamic structure on the front zone of a cylinder in crossflow of water at $Re_f = 1 \cdot 10^4$.

basis of a number of accurate measurements [107, 108]—by hydrogen dye techniques, by thermoanemometry, and from correlation in a wide range of Re (up to $5 \cdot 10^4$).

The stability of the cellular structure has not yet been studied. In our opinion, the stability must be related to the free-stream velocity, the surface curvature, and the boundary-layer parameters. On the basis of an analysis of the cellular structure and of its stability, it might be possible to explain the early transition in the boundary layer that occurs for high levels of turbulence and for critical values of Re.

5.2 THE DETERMINING FACTORS OF HEAT TRANSFER AT THE FRONT STAGNATION POINT

Heat transfer at the front stagnation point is influenced most strongly by the level of turbulence in the free stream [18, 114]. The thickness of the boundary layer is constant on the front part of the cylinder and depends only on Re. At higher angular distances φ, the effect is less pronounced because the thickness of the boundary layer increases and the penetration of the external turbulence decreases, which is reflected in the heat transfer. The cellular structure in the vicinity of the front stagnation point may be a further factor for the augmentation of the heat transfer. Numerous studies have shown that an increase in turbulence to 3% is accompanied by a 60% or more increase of the heat transfer coefficient at the front stagnation point. The effect of the turbulence length scale is as yet obscure. In fact, there is some disagreement about the general effect of the turbulence level. Some authors [105] maintain that the heat transfer does not respond to turbulence of low levels ($Tu \approx 1\%$). Others present data that appear to demonstrate that the heat transfer coefficient increases continuously with turbulence level. None of the published studies deals adequately with turbulence effects for a wide range of fluids, neither at the front stagnation point nor in the other circumferential regions.

The measured values of shear stress and of pressure distribution (Chap. 4) suggest that there is no effect, over the range of flow regimes and geometries studied, of the free-stream turbulence on the shear stress near the front stagnation point. An effect is only observed beyond a certain angular distance, typically $\varphi \approx 20°$. This observation contradicts the Reynolds analogy, which suggests that friction drag and thus shear stress should increase in parallel with the increase of the heat transfer. The contradiction must be related to the vortex processes in the vicinity of the front stagnation point.

In terms of the theory of similarity, the most general description of heat transfer at the front stagnation point is

$$\mathrm{Nu} = f(\mathrm{Re},\ \mathrm{Pr},\ Tu,\ k_q,\ L,\ \Lambda) \tag{5.23}$$

There exist a number of empirical relations for the heat transfer. They are based either on the theory of similarity, or on other assumptions, such as the analogy between the decay of turbulence in the boundary layer and in the viscous sublayer. The data are most often interpreted in the forms

$$\frac{\mathrm{Nu}_{Tu}}{\mathrm{Nu}_{Tu=0}} = f(\mathrm{Re}_t) = f(Tu\sqrt{\mathrm{Re}}) = f(\sqrt{Tu\,\mathrm{Re}}) \tag{5.24}$$

Two-term equations of this nature were suggested by Dyban et al. [110], who suggested the equations

$$\frac{\mathrm{Nu}_{Tu}}{\mathrm{Nu}_{Tu=0}} = 1 + 0.01\sqrt{\mathrm{Re}\,Tu} \tag{5.25}$$

or

$$\frac{\mathrm{Nu}_{Tu}}{\mathrm{Nu}_{Tu=0}} = 1 + \frac{0.8\,\mathrm{Re}\,Tu}{1500 + \mathrm{Re}\,Tu} \tag{5.26}$$

Alternative three-term expressions are given by Kestin [33] and by other authors.

On the basis of Eq. (5.23) we propose a power law relation for heat transfer at the front stagnation point, of the following form:

$$\mathrm{Nu}_f = c\,\mathrm{Re}_f^m\,\mathrm{Pr}_f^n\,Tu^k\,(\mathrm{Pr}_f/\mathrm{Pr}_w)^{0.25} \tag{5.27}$$

Let us consider first heat transfer in the subcritical flow regime. The analysis in Section 5.1 suggests the exponent on the Prandtl number should be 0.35 to take proper account of the fluid physical property effects. This analytical result is consistent with experiments in air, water, and transformer oil (Figs. 5.5 and 5.6). Figure 5.5 demonstrates the significant increase in heat transfer coefficient with an increase in free-stream turbulence. Furthermore, it is evident from Fig. 5.5 that the exponent on Re_f increases with increasing free-stream turbulence. For $Tu = 1\%$, the exponent on Re is practically constant at $n = 0.5$, and coincides with the analytical prediction [18, 19]. Our studies carried out in collaboration with Katinas [74] and the results of Dyban et al. [110] show that the exponent n on Re_f increases to 0.6 for an increase of Tu up to 7 or 8%. The variation of n was also confirmed in the studies with air [110, 25]. We tenta-

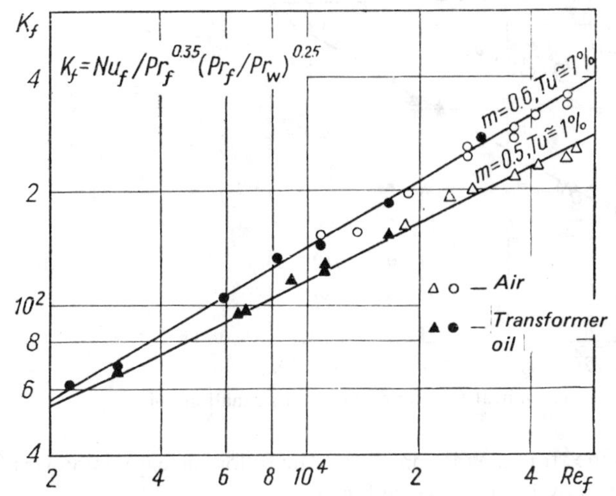

FIG. 5.5 Heat transfer at the front stagnation point in the subcritical range of Re.

tively ascribe this variation to the pseudo-laminar nature of the boundary layer and to the spatial structure of the fluid dynamics in the vicinity of the front stagnation point.

The results for the lower level of turbulence ($Tu \leqslant 1.0\%$) in Fig. 5.6 are approximated by

$$Nu_f = 1{,}11\ Re_f^{0.5}\ Pr_f^{0.35}\ (Pr_f/Pr_w)^{0.25} \tag{5.28}$$

and agree well with the analytical solution (solid line) by Eckert [19]. This solution applies to the heat transfer at the front stagnation point for $Tu = 0$ or for a low level of turbulence. We performed experiments with a wide range of fluids and free-stream turbulence, in order to evaluate quantitatively an expres-

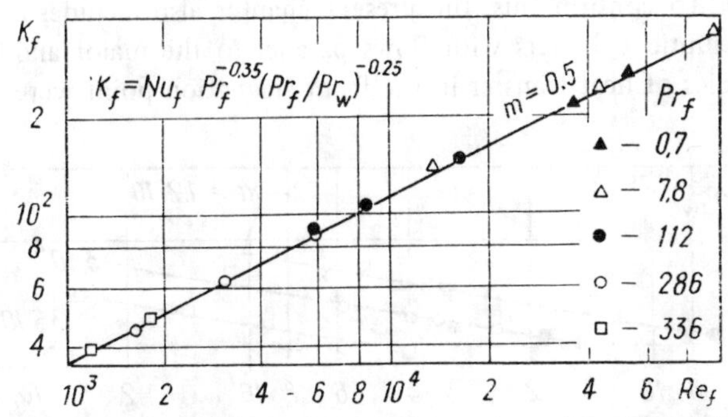

FIG. 5.6 Heat transfer at the front stagnation point with low level of free-stream turbulence ($Tu = 1.0\%$).

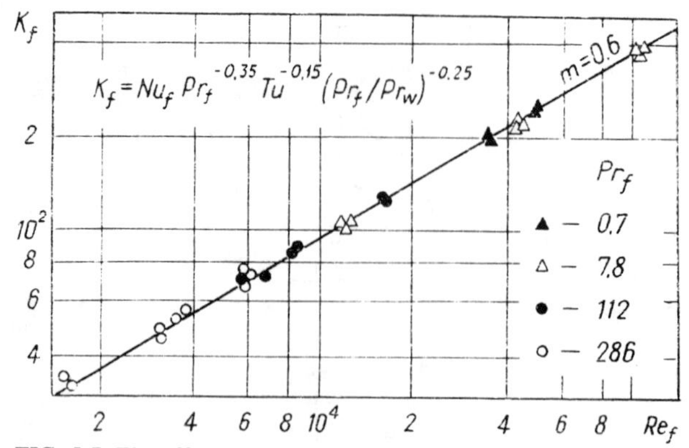

FIG. 5.7 The effect of turbulence on heat transfer at the front stagnation point.

sion for heat transfer in the front stagnation point. The following expression was obtained, in the form given by Eq. (5.27) (see Fig. 5.7):

$$\mathrm{Nu}_f = 0.41\, \mathrm{Re}_f^{0.6}\, \mathrm{Pr}_f^{0.35}\, Tu^{0.15}\, (\mathrm{Pr}_f/\mathrm{Pr}_w)^{0.25} \tag{5.29}$$

Here, for high levels of turbulence the power index of Re_f is 0.6. An evaluation of our results and the results by other authors (Fig. 5.8) shows that there is a threshold value of the turbulence level, where the effect of turbulence on heat transfer is initiated. The threshold is lower at higher Re_f; thus $Tu \approx 1\%$ at $\mathrm{Re} = 1 \cdot 10^5$, and $Tu \approx 5\%$ at $\mathrm{Re}_f = 3.5 \cdot 10^3$. The phenomenon is mostly probably related to changes in the spatial three-dimensional cellular structure of the fluid dynamics.

The fluid dynamic studies on cylinders of different cross sections (see Chap. 4) showed that there was close dependence between shear-stress distribution and the closeness of the geometry to that of a slender body. The heat transfer coefficient in the various regions of a cylinder, as well as its average value, must also depend on the geometry and on the yaw angle in which the cylinder is mounted in the channel. To confirm this, the present chapter also includes a description studies on elliptic cylinders with flows parallel to the major and the minor axes. The studies of heat transfer in the front stagnation point were

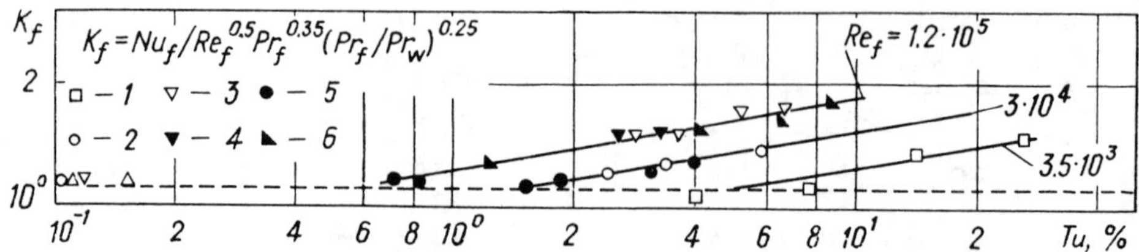

FIG. 5.8 Heat transfer in the front stagnation point for different Re_f and Tu: 1, our results for air; 2, 3, 4, from [32]; 5, from [35]; 6, our results for water.

performed with turbulence level below 1.0%, so that the only factors to be accounted for were the physical properties and the geometry. Fluid physical properties were represented by Pr to the 0.33 power, and free-stream velocity effects were represented by an exponent of 0.5 on Re_f (Figs. 5.9 and 5.10).

For the elliptic cylinder, with the flow parallel to the major axis, the following expression was obtained for Nu_{fd_1}

$$\mathrm{Nu}_{fd_1} = 1.65\ \mathrm{Re}_{fd_1}^{0.5}\ \mathrm{Pr}_f^{0.33}\ (\mathrm{Pr}_f/\mathrm{Pr}_w)^{0.25} \tag{5.30}$$

Similarly, with the flow parallel to the minor axis, Nu_{fd_2} is given by

$$\mathrm{Nu}_{fd_2} = 0.75\ \mathrm{Re}_{fd_2}^{0.5}\ \mathrm{Pr}_f^{0.33}\ (\mathrm{Pr}_f/\mathrm{Pr}_w)^{0.25} \tag{5.31}$$

The parameter values in Eqs. (5.30) and (5.31) were referred to the corresponding axis of the cylinder.

Our studies carried out in collaboration with Katinas [111] showed that the more slender the geometry, the higher the heat transfer coefficient at the front stagnation point. The data shown in Figs. 5.9 and 5.10 show somewhat higher values of heat transfer coefficient than in the earlier work, probably due to a somewhat higher free-stream turbulence.

The heat transfer near the front stagnation point is related to the flow behavior around the cylinder and in the near wake. Our studies on different cylinders in a wide range of Re_f showed that the effect of turbulence was different in the subcritical and in the critical ranges of Re_f.

The results obtained by Ilgarubis et al. [38] for the critical flow regime show that the heat transfer coefficient increases with turbulence level up to $Tu = 3$–4% but the further increase of turbulence beyond this level has no further effect on heat transfer (Figs. 5.11 and 5.12). For subcritical flow

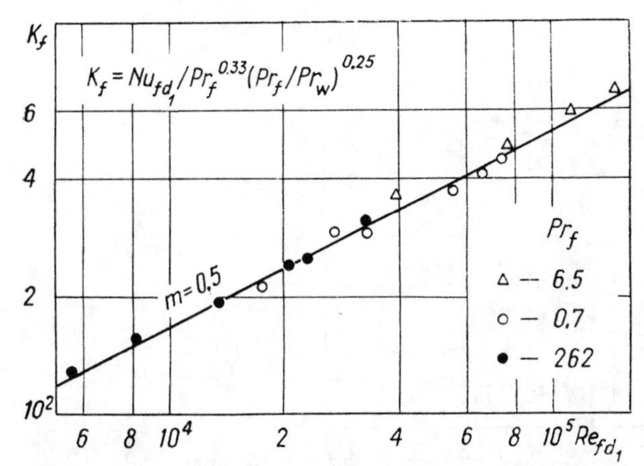

FIG. 5.9 Heat transfer at the front stagnation point on an elliptic cylinder with the flow parallel to the major axis.

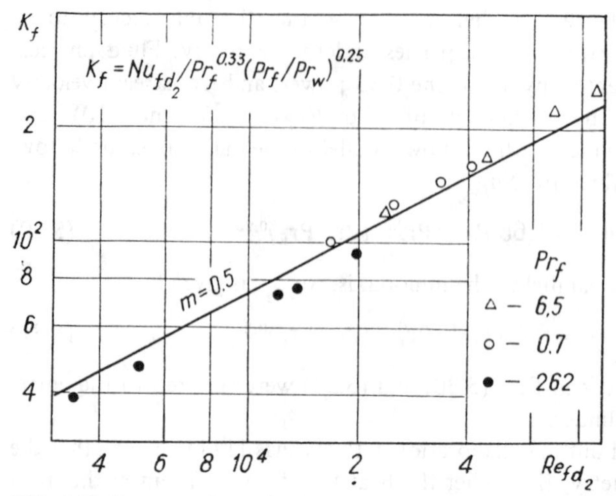

FIG. 5.10 Heat transfer at the front stagnation point on an elliptic cylinder with the flow parallel to the minor axis.

($Re_f \leqslant 10^5$), a continuous augmentation of the heat transfer is noted with an increasing turbulence level (Fig. 5.11). In the critical flow regime there is a very thin laminar boundary layer in the vicinity of the front stagnation point, and this is easily penetrated even by moderate turbulent fluctuations. In the higher levels of turbulence, the fluctuations are more powerful but exert insignificant further effect on the heat transfer, at least in the range of turbulence studied. On the other hand, as Re_f is further increased, a finer cellular structure is observed for the turbulence and thus can no longer penetrate the boundary layer

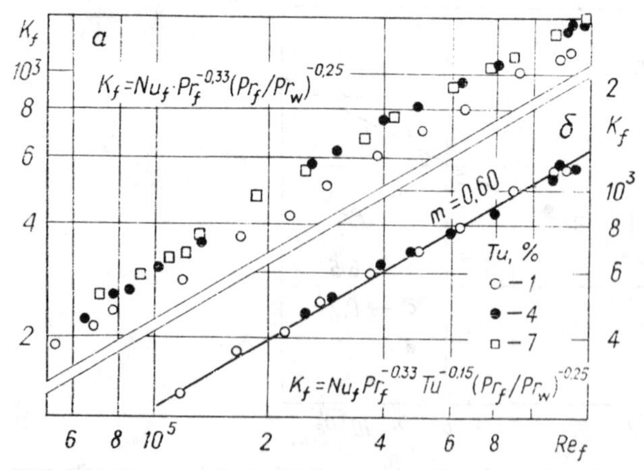

FIG. 5.11 Heat transfer in the front stagnation point of a cylinder in water in the critical range of Re_f: Expressions for K_f calculated not including Tu (α) and including Tu (δ).

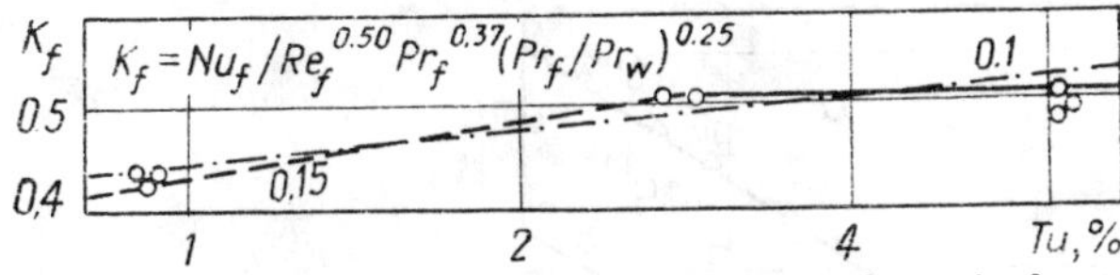

FIG. 5.12 The effect of turbulence on heat transfer at the front stagnation point in the critical range of Re_f.

and increase the heat transfer coefficient. A similar tendency was noted in the gradient flow region adjacent to the stagnation region: Kestin et al. [109] report that free-stream turbulence has a negligible effect on a turbulent (and, we suggest a pseudo-laminar) boundary layer.

To represent the experimental results on the effect of the free-stream turbulence (Fig. 5.11, δ) the term $Tu^{0.15}$ was introduced into Eq. (5.32). Thus, for design calculations we propose the expression

$$Nu_f = 0.326\, Re_f^{0.6}\, Pr_f^{0.33}\, Tu^{0.15}\, (Pr_f/Pr_w)^{0.25} \tag{5.32}$$

for the critical flow regime and for low levels of turbulence. Note that the exponent on Re_f in Eq. (5.32) is 0.6, i.e., the same as for subcritical flow, due to the pseudo-laminar nature of the boundary layer.

To cover the full range of *Tu* studied, an approximate exponent of 0.1 may be used on *Tu* (see Fig. 5.12). When this average exponent is used in Eq. (5.32), an underestimate of heat transfer coefficient results for the lower levels of turbulence, and an overestimate for the higher ones.

In air, the effect of turbulence is continuous, similar to the behavior observed in the subcritical flow regime in other fluids (Fig. 5.13). A comparison (Fig. 5.14) of the results for water and for air [37] revealed that the effect of turbulence in air was more pronounced. It was also observed that the Nusselt numbers were similar for the two flows with low levels of turbulence, the differences only being observed with higher levels of turbulence. Although studies of this effect are continuing, we may tentatively conclude that the difference is due to differences in the interaction between the turbulent fluctuations and the thermal boundary layer. The thermal boundary layer is much thinner in water since, in fluids with higher Pr_f, the thermal resistance is concentrated near the wall. Thus, the heat transfer coefficient is less strongly affected by the turbulence and the velocity fluctuations. An analysis of the structure of the turbulent boundary layer indicates a strong sensitivity of heat transfer coefficient to the thickness of the viscous sublayer. Thus, the effect of the free-stream turbulence must be lower in fluids with a high Pr_f.

The results shown in Fig. 5.13 suggest the exponent on *Tu* is around 0.2 for air, resulting in the following expression for Nu_f:

$$Nu_f = 0.32\, Re_f^{0.6}\, Tu^{0.2} \tag{5.33}$$

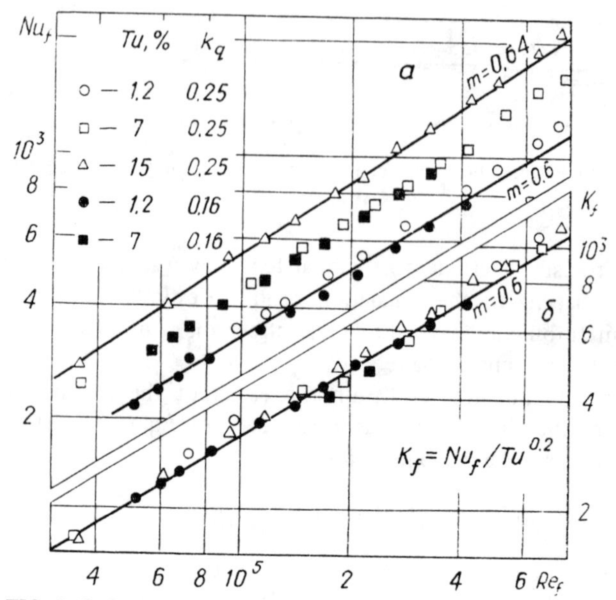

FIG. 5.13 Heat transfer at the front stagnation point for flow over a cylinder at high Re_f in air: Nu_f as a function of Re_f (α) and $K_f = Nu_f/Tu^{0.2}$ as a function of Re_f (δ).

The results for air show that the value of Nu_f is practically independent of the cylinder diameter or of the blockage factor, except for the extreme values of the latter. On the other hand, the results for water show a strong dependence on the blockage factor (Fig. 5.15). Our study included two independent procedures for changing the blockage ratio: k_q was varied from 0.28 to 0.70 either by

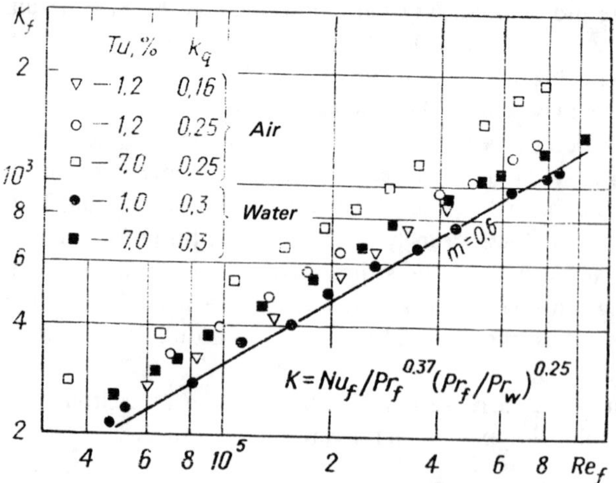

FIG. 5.14 Comparison of heat transfer behavior at the front stagnation point for a cylinder in water and in air, respectively.

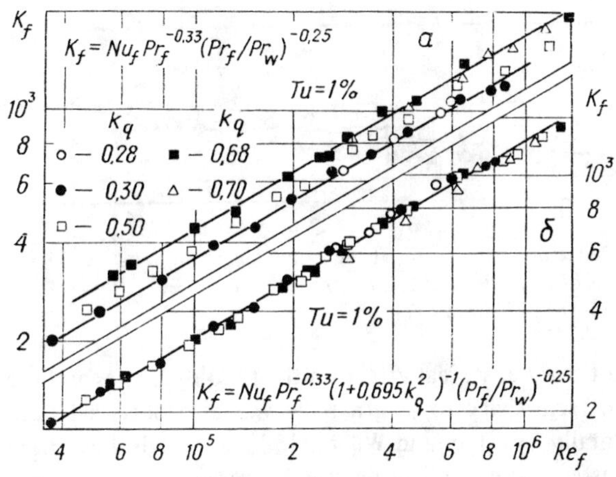

FIG. 5.15 Heat transfer behavior at the front stagnation point of a cylinder in the high range of Re_f: (α) K_f not including blockage factor k_q; (δ) K_f including k_q.

changing the cylinder diameter with a constant channel height, or vice versa. Irrespective of the procedure, the heat transfer coefficient at the front stagnation point increased with increasing k_q, as represented by the equation

$$Nu_f = 0.326\, Re_f^{0.6}\, Pr_f^{0.33}\, (1 + 0.7\, k_q^2)^{-1}\, (Pr_f/Pr_w)^{0.25} \tag{5.34}$$

which is valid only for $Tu \leqslant 1\%$. The factor $(1 + 0.7k_q^2)^{-1}$ is identical to that suggested earlier by Akylbaev et al. [39] for the subcritical flow regime. The analysis of fluid dynamics described in Chap. 4 indicates that the variations of the heat transfer coefficient at the front stagnation point, due to changes in blockage factor, are mainly governed by the changes in the pressure and velocity gradients.

In addition to the above studies of turbulence whose scale was small compared to the cylinder diameter, we also studied heat transfer at the front stagnation point in turbulent flows of much longer scale. The results (illustrated in Fig. 5.16) show that heat transfer was not enhanced over the range of length scale (L) studied. Indeed, an adverse effect of the macroscale turbulence appears with $(L/d)\, Re_f^{0.6} \geqslant 10^3$. The size of the vortices investigated is close to that of

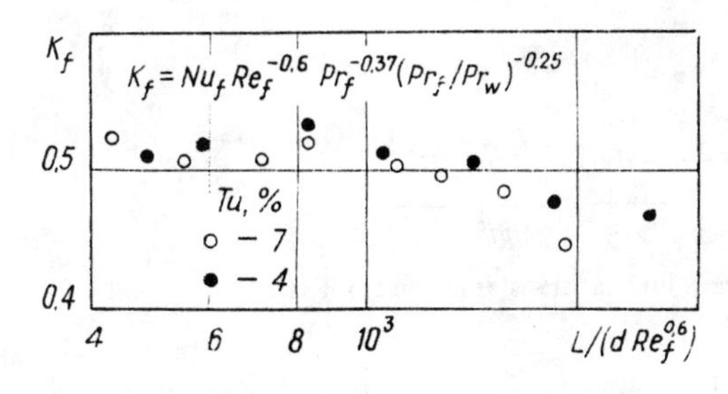

FIG. 5.16 The effect of the integral length scale of turbulence on heat transfer at the front stagnation point.

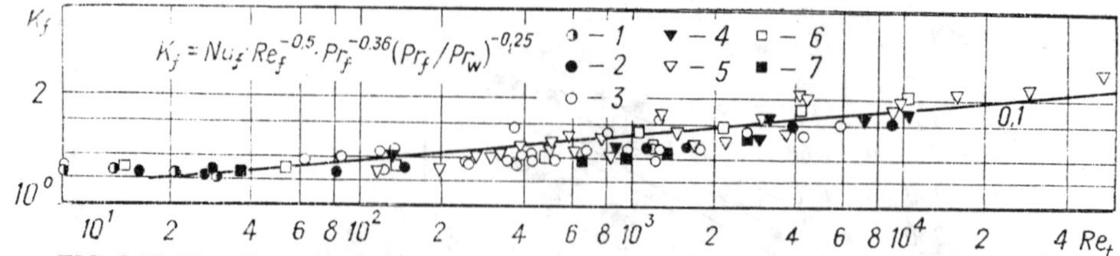

FIG. 5.17 The effect of turbulence on heat transfer at the front stagnation point. A modified interpretation in terms of $Re_t = Re\,Tu$. 1, $Tu = 0.5\%$; 2, $Tu = 1\%$; 3, $Tu = 4\%$; 4, $Tu = 6\%$; 5, $Tu = 9\%$; 6, from [85]; 7, from [37].

the cylinder, so they do not exert any great effect on the fluid dynamics and the heat transfer. This lack of sensitivity to length scale was also observed both analytically and experimentally by Traci and Wilcox [58], although they noted an increase of the heat transfer as the length scale was decreased.

The equations given above that contain Tu raised to a fixed power exhibit uncertainty in the lower levels of turbulence (Tu from 0 to 1.0%). However, we maintain that for most common practical implementations employing turbulent flows of air and water, the approximations are sufficiently accurate. To compare our results with the earlier publications, and to attempt to obtain a higher degree of generality, we also interpreted the data in terms of a modified value of the Reynolds number $Re_t = \sqrt{\overline{u'^2}}\,d/\nu = Re\,Tu$, which is referred to the absolute value of the velocity fluctuation.

The parameter K_f (Fig. 5.17) was plotted in terms of Re_t, and the present and literature data were well represented by a power law of the form $K_f \sim Re_t^{0.1}$. Thus, Nu_f is

$$Nu_f = 0.73\,Re_f^{0.5}\,Pr_f^{0.36}\,Re_t^{0.1}\,(Pr_f/Pr_w)^{0.26} \tag{5.35}$$

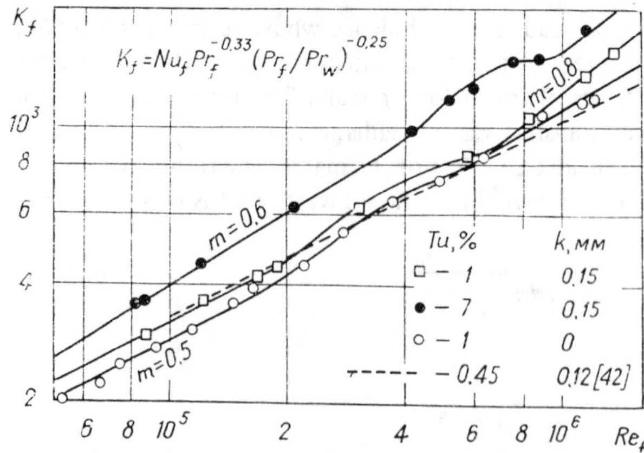

FIG. 5.18 The effect of surface roughness on heat transfer at the front stagnation point on a cylinder.

Results were also obtained (Fig. 5.18) for the effect of surface roughness on heat transfer at the front stagnation point. The heat transfer coefficient shows a general tendency to increase with roughness, the specific behavior depending on the level of turbulence. At lower levels of turbulence, Nu_f varies with $Re_f^{0.5}$ and at higher levels ($Tu \geqslant 4\%$) with $Re_f^{0.6}$. For $Re_f > 10^6$, Nu_f varies with $Re_f^{0.8}$. Good agreement is obtained with the results of other authors [42].

CHAPTER

SIX

LOCAL HEAT TRANSFER

The highly complicated structure of the fluid dynamics leads to the existence of various regions on the cylinder surface. The laminar boundary layer that exists on the front part can become partially turbulent under certain conditions. Laminar-turbulent transition in the boundary layer and boundary-layer separation are governed by Re. In the rear of the cylinder, vortex flow occurs. The two stagnation points (one in the front and one in the rear) must be treated separately because of their specific fluid dynamics. It is only reasonable to expect that heat transfer on the cylinder surface will be of a similarly complex character. Any study of heat transfer in general and of its specific features begins quite naturally with determination of the local heat transfer coefficient. Knowledge of the local heat transfer coefficient around a cylinder, particularly under high thermal loads, allows prediction of the temperature distribution (and hence of zones subject to overheating) on the one hand, and the ways of augmenting the heat transfer on the other. With this in mind, we included in our program a study of the effects, on the local heat transfer coefficients, of the free-stream turbulence, the blockage factor, the surface roughness, the fluid physical properties, and the fluid flow regime. Ranges of Re_f from 10 to $2 \cdot 10^6$ and of Pr_f from 0.7 to 1000 were covered.

6.1 THE CIRCULAR CYLINDER IN THE SUBCRITICAL FLOW REGIME

On the front part of a cylinder in crossflow, where a laminar boundary layer develops, the heat transfer coefficient can be determined by either an approximate or an exact numerical analysis. The various solutions are illustrated in Fig. 6.1. The approximate solution by Kruzhilin [17] and Frössling [18] were based on solutions of the momentum and energy equations using polynomial descriptions of the velocity distribution in the boundary layer. The techniques used by Eckert [19, 104] and Merk [112] were based on the boundary-layer equations and could give predictions of local heat transfer coefficient only up to the separation point (Fig. 6.1). Our analysis [101] (for $Re_f = 4.96 \cdot 10^4$), described in Chap. 3, is in good agreement with the other results shown in Fig. 6.1. The deviation that occurs as the separation point is approached is most probably due to the different velocity profiles in the boundary layer and different velocity distributions outside the boundary layer assumed in the various studies. The curves in Fig. 6.1 cover only angular distances up to $\varphi \approx 70°$. None of the known analytical techniques applies for the whole cylinder surface over a wide range of Re. The available studies (such as those by Krall and Eckert [104]) apply only to low ranges of Re. In these studies, local heat transfer coefficients were determined for both T_w = constant and q_w = constant boundary conditions from numerical solutions of the boundary-layer equations (Fig. 6.2). Constant heat flux gives a higher heat transfer coefficient over the larger part of the circumference. On the front part of the cylinder (up to $\varphi \approx 25$–$30°$), there is no appreciable effect of boundary condition. In the flows in this region the temperature profiles, and consequently the heat transfer coefficients, are insensitive to the boundary conditions on the surface. Higher heat transfer coefficients have also been observed experimentally with constant heat flux (q_w = constant), in both subcritical and critical flow around cylinders, and on a plate with a laminar boundary layer [121]. The predicted trends for local heat

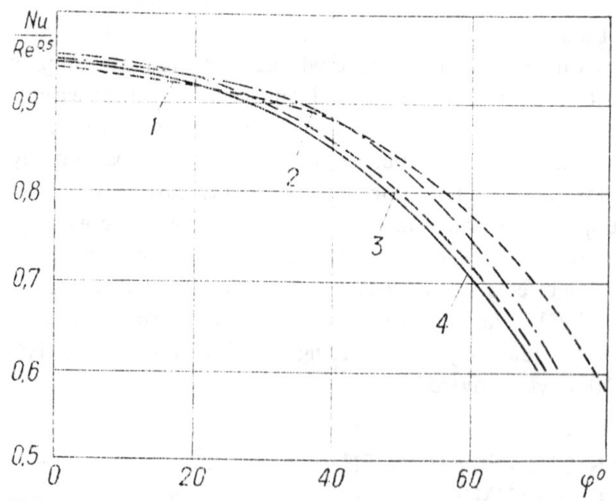

FIG. 6.1 Analytical curves for local heat transfer in the front part of a cylinder: 1, [17]; 2, [101]; 3, [19]; 4, [18].

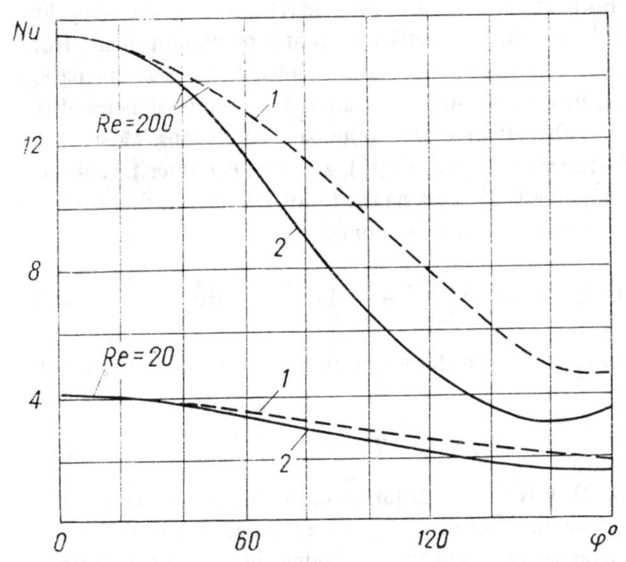

FIG. 6.2 Local heat transfer in flow over a cylinder with alternative boundary conditions for low Re: 1, q_w constant; 2, T_w constant.

transfer coefficient shown in Fig. 6.2 were confirmed experimentally by Eckert [104], and the new results shown in Fig. 6.3 for crossflow of aviation oil ($\mathrm{Pr}_f = 1000$) over a cylinder. The main discrepancies are in the prediction of the position of the minimum heat transfer coefficient, which occurs at $\varphi \approx 120°$, in contrast to the predicted value of $\varphi \approx 150°$, and also in the shape of the lowest

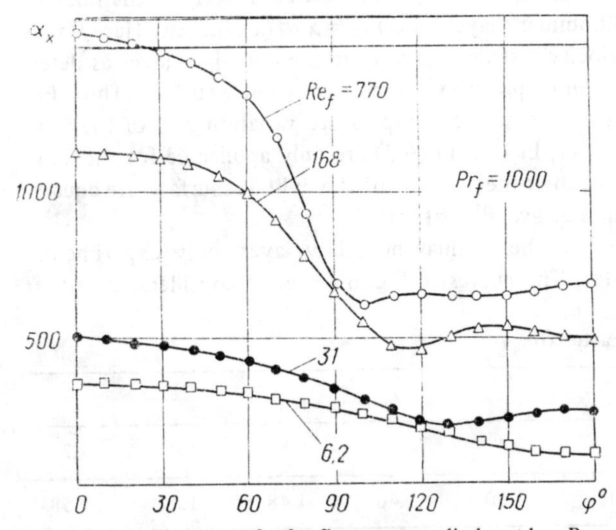

FIG. 6.3 Local heat transfer for flow over a cylinder at low Re_f.

curve for $Re_f = 6.2$. Both experimentally and analytically, at the very low Reynolds numbers, the heat transfer coefficient is at a maximum in the front zone of the cylinder, and at a minimum in the vicinity of the rear stagnation point. The results shown in Figs. 6.1-6.3 illustrate the general tendencies of the local heat transfer and of the effect of the boundary conditions. Most of the earlier analytical publications (e.g., [17]-[19]), and Fig. 6.1 refer to constant surface temperatures. Frössling [18] suggests the following relationship for local heat transfer coefficient up to the separation point:

$$Nu_x = \left[0.945 - 0.7696\left(\frac{x}{d}\right)^2 - 0.3478\left(\frac{x}{d}\right)^4\right] Re_x^{0.5} \tag{6.1}$$

For constant wall heat flux, the wall temperature at the front zone could be represented in the form:

$$T_w - T_f = \Delta t = T_0\, x^{n_0} \tag{6.2}$$

In Eqs. (6.1) and (6.2), x is the arc distance on the circumference from the front stagnation point. We developed [113] an extended solution for heat transfer near the front stagnation point by employing the power-law temperature distribution given by Eq. (6.2) and a power-law velocity distribution outside the boundary layer ($U = U_\infty x^{m_0}$) using a technique that was a modified form of that suggested by Eckert [19]. The expression is

$$Nu_x = 0.332\, f(m_0\, n_0)\, Re_x^{0.5}\, Pr^{0.333 + 0.067\,\beta - 0.026\,\beta} \tag{6.3}$$

where $f(m_0 n_0) = x(m_0)\psi_1(n_0) + \psi_2(n_0)$. Values of the factor $f(m_0 n_0) = x(m_0)\psi_1(n_0) + \psi_2(n_0)$ are given in Tables 6.1 and 6.2 for different values of m_0 and n_0. The parameter $\beta = 2m_0/(m_0 + 1)$ is related directly to the velocity distribution outside the boundary layer, and $m_0 = x\, df(x)/f(x)\, dx$. Here $f(x)$ is a function giving the velocity distribution outside the boundary layer as determined from the distribution of pressure around the cylinder surface. Thus, Eq. (6.3) predicts the effect of the surface-temperature variation and of the fluid physical properties. However, Eqs. (6.1)-(6.3) are only applicable for the laminar boundary-layer region; the other regions of the cylinder surface are equally important in the evaluation of overall heat transfer.

For the regions beyond the laminar boundary layer, only experimental results are available so far. The curves in Fig. 6.4 suggest considerable further

TABLE 6.1 Values of the Factor $x(m_0)$

	m_0						
	0	0.2	0.4	0.6	0.8	1.0	1.2
$x(m_0)$	0.92	1.15	1.30	1.40	1.48	1.54	1.58

TABLE 6.2 Values of the Factors $\psi_1(n_0)$ and $\psi_2(n_0)$

	n_0									
	−0.5	−0.4	−0.2	0.0	0.2	0.4	0.6	0.8	1.0	1.2
$\psi_1(n_0)$	1.6	1.28	1.03	0.94	0.96	0.96	0.94	1.0	1.08	1.15
$\psi_2(n_0)$	−1.6	−0.93	−0.3	1.0	0.23	0.34	0.39	0.41	0.43	0.44

changes in these regions of the local heat transfer coefficient at higher Re_f [111]. At lower Re_f (Figs. 6.2 and 6.3), the boundary-layer separation is located in the rear zone at $\varphi = 125$–$150°$. At higher Re_f, it is shifted to $\varphi = 82$ or $90°$, and remains at this location until the critical flow regime is established, $\mathrm{Re}_f \approx 2 \cdot 10^5$. As will be seen in Fig. 6.4, at lower Re_f the heat transfer coefficient is approximately the same at the two stagnation points. Note that the heat transfer coefficient for water (Pr = 7.5) is higher than for the other fluids. At lower Re_f (up to $\mathrm{Re}_f = 5 \cdot 10^3$), the average heat transfer coefficient is higher in the front half of the cylinder than it is in the rear (see Figs. 6.2 and 6.3).

Local heat transfer is also affected by free-stream turbulence. In our studies in the subcritical flow regime, this effect was found to be much more pronounced than that of surface boundary conditions. In Fig. 6.2, the effect of these boundary conditions is shown; this effect is different in the various regions

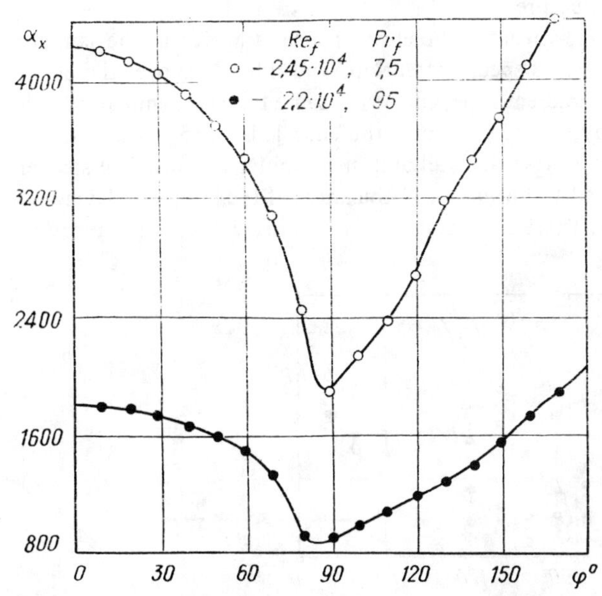

FIG. 6.4 Local heat transfer coefficients for flows of eater and transformer oil over a cylinder.

of the surface. In general, with a constant heat flux, the heat transfer coefficient is 10 to 20% higher than with a constant surface temperature. The difference varies with Re_f, but never exceeds 20%. In contrast, our studies in turbulent flows of different fluids (Fig. 6.5), performed in collaboration with Katinas [111], revealed a 30–50% and even higher increase of the local heat transfer coefficient in the various surface regions as the turbulence level was increased. As is clear from Fig. 6.5, the effect of turbulence is also localized and is strongest in the front part of the cylinder, the effect becoming less pronounced at higher angular distances. No systematic effect of turbulence can be observed in the rear zone. For $Re_f = 1.85 \cdot 10^4$ the turbulence effect decreases continuously as the rear stagnation point is approached, and completely disappears at the stagnation point (Fig. 6.5). At $Re_f = 4.96 \cdot 10^4$ and $Tu = 7.2\%$, a sharp peak in heat transfer coefficient is observed in the intermediate part of the circumference. Here, the curve has two minima, which is a feature of the critical flow regime. For the details of heat transfer in this regime, we refer the reader to Section 6.3 and only note here that a 15% decrease of the heat transfer coefficient occurs in the rear zone, where turbulence is increased from 0.5 to 7.2%. The decrease is related to the breakdown of a regular vortex shedding and to the narrowing of the near wake, due to the downstream shift of the separation point. On the front zone, the boundary-layer thickness increases with the angular distance, and the effect of the turbulence level decreases. In the rear part, the effect is insignificant, because the separation of the boundary layer is followed by a highly turbulent vortex structure.

The effect of the free-stream turbulence on heat transfer in the laminar boundary-layer region deserves special attention. The fluid flow over the cylinder in crossflow is of the gradient type, and at the same time the laminar boundary layer is influenced by the free-stream turbulence [114, 115]. The velocity profiles are deformed both near the wall and in the outer regions. The steeper velocity profiles are related to an increased thickness of the boundary layer and a higher velocity gradient. These are accompanied in turn by steeper temperature

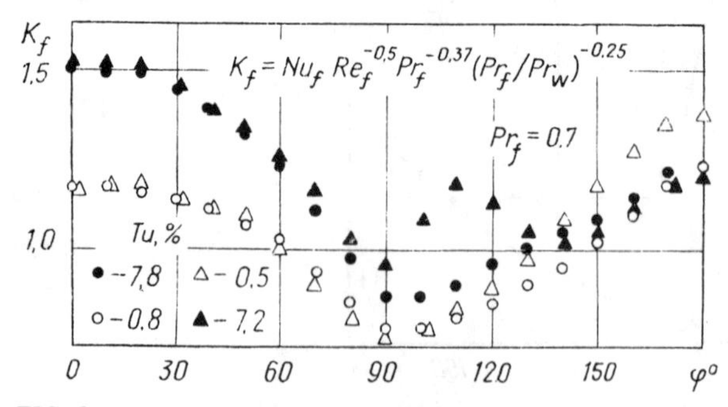

FIG. 6.5 Local heat transfer coefficients for air over a cylinder at different turbulence levels.

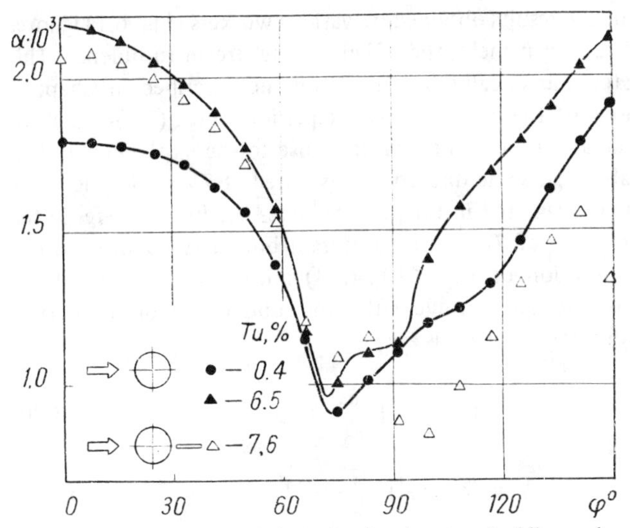

FIG. 6.6 The effect of turbulence level and vortex shedding on local heat transfer coefficient.

gradients at the wall, and by a thicker thermal boundary layer; the steeper gradients mean that the heat transfer coefficient is also higher. Furthermore, the cellular flow structure, described in Chap. 5, may also contribute to the heat transfer.

Regular vortex flow in the rear part of a cylinder has been observed by numerous authors in the subcritical flow regime. The vortices are shed intermittently from the two sides of the cylinder and generate fluctuations of the velocity in the front part. In order to establish the magnitude of the effect of the vortices on heat transfer, it was necessary to carry out tests in which regular vortex shedding is eliminated. This was done by the established technique of placing a partitioning splitter plate in the near wake. The 3-mm-thick plate was five cylinder diameters long and was located at 1 mm from the rear stagnation line.

In the subcritical flow regime, the elimination of vortex shedding leads to a small decrease of the heat transfer in the front zone (Fig. 6.6). The effect depends on Re_f and Tu, the decrease being more pronounced at lower Re_f. A sharp decrease in heat transfer coefficient was noted in the rear zone. A further minimum appeared in the curve of the local heat transfer coefficient, and this minimum moved upstream with increasing Tu.

The fluctuations of velocity resulting from the vortex shedding are seen to have little effect on the heat transfer in the front part, but their elimination leads to a considerable decrease of the heat transfer in the rear part of the cylinder since the macroscale vortex transfer in the near wake is obviously changed.

The comparison of the results obtained by various workers (Fig. 6.1) ignores one highly important factor–namely, the effect of free-stream turbulence. This effect can be estimated numerically by the technique described in Chap. 3. The numerical solutions of the boundary-layer equations, Eqs. (3.1)-(3.3), for subcritical flows of air and transformer oil are close to the experimental data (Figs. 6.7 and 6.8), although some discrepancy is noted at $Tu = 4\%$ (Fig. 6.7). The analytical technique suggested in Chap. 3 evaluates the heat transfer in the front part for a wide range of Tu and for different boundary conditions. For turbulent flows, any solution of Eqs. (3.1)-(3.3) must take account of the effects of turbulent viscosity and turbulent thermal conductivity on shear stress and heat flux. The appropriate equations are:

$$\tau = (\mu + \mu_{Tu}) \frac{\partial u}{\partial y} \tag{3.20}$$

$$q = (\lambda + \lambda_{Tu}) \frac{\partial T}{\partial y} \tag{3.21}$$

The following relation for turbulent viscosity and turbulent thermal conductivity was used:

$$\frac{\lambda_{Tu}}{c_p} = \frac{\mu_{Tu}}{\mathrm{Pr}_t} \tag{3.22}$$

where

$$\mu_{Tu} = K \frac{Tu}{100} \delta U_\infty n^* \tag{3.23}$$

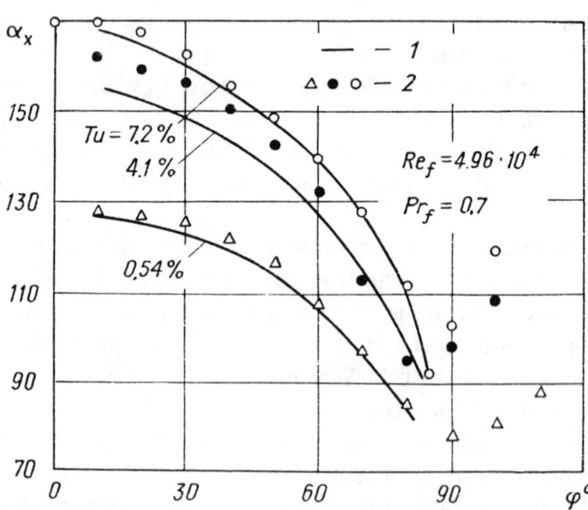

FIG. 6.7 Comparison of predicted and measured local heat transfer coefficients for air flow: 1, analysis; 2, experiment.

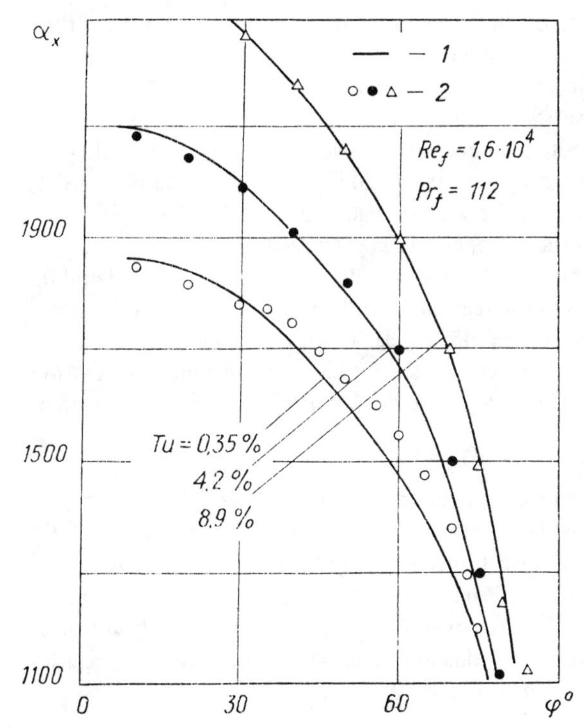

FIG. 6.8 Comparison of predicted and measured local heat transfer coefficients in transformer oil flow: 1, analysis; 2, experiment.

Further details are given in Section 3.4. The technique yields heat transfer coefficients for angular distances from $\varphi = 2°$ to the separation at $\varphi \approx 80°$, but it does not apply to the front stagnation point itself.

Figures 6.3–6.6 present general curves for the local heat transfer coefficient around the whole circumference for the subcritical flow regime, and the separate fluid dynamic regions are easily distinguished. The characteristics of the curves are similar in different fluids, any differences being related to either the onset of the critical flow regime, or the elimination of the regular vortex shedding.

In engineering design, the local heat transfer behavior is often of subsidiary interest, and general overall average relationships like

$$\mathrm{Nu} = c\,\mathrm{Re}^m\,\mathrm{Pr}^n \tag{1.1}$$

are often used. Since relationships result from integration of the local heat transfer coefficient over part or all of the cylinder surface. First, we will consider separately the heat transfer in the front half and in the rear half of the cylinder, respectively. The onset of a specific flow regime should be reflected in equations of the form of Eq. (1.1) by a change in the value of the power index. Numerous results of our studies and of [18] suggest $m = 0.5$ as the power index on Re_f for the laminar boundary layer on the front part of the cylinder. However, this is only true for lower levels of turbulence ($Tu < 1\%$). An increase in turbulence involves an increase of the exponent; thus, $m = 0.65$ at $Tu > 20\%$. For the rear part, the value of m varies from 0.5 to 0.8, and an average value $m = 0.73$ may be assumed throughout. In the studies carried out by Dyban et al. [110] in air at $Tu > 12\%$, a practically constant value of $m = 0.66$ was observed for the whole circumference and for the whole range of Tu from 12 to 23%.

The effect of fluid physical properties, as expressed in terms of Pr_f^n, is more uniform over the circumference. The power index n has values of 0.34 and 0.4, respectively, for the front and the rear parts, and a value of 0.3 at the separation point at $\varphi = 80°$. The value of n did not appear to vary with Tu.

Analysis of our results (Fig. 6.9) indicates that, for constant wall heat flux, the average heat transfer coefficient for the front half of the cylinder is given by:

$$\mathrm{Nu}_{fx} = 0.65\,\mathrm{Re}_{fx}^{0.5}\,\mathrm{Pr}_f^{0.33}\,(\mathrm{Pr}_f/\mathrm{Pr}_w)^{0.25} \tag{6.4}$$

Here the dimensionless groups are calculated in terms of the local velocity given by Eq. (3.6) and of the angular distance from the front stagnation point. With constant wall temperature the heat transfer coefficient is 15 to 20% lower.

The effect of the turbulence level on heat transfer between a cylinder and a crossflow of air was studied by Dyban et al. [110]. Our results for heat transfer in the front zone for a range of fluids are described by the following relation, which takes account of turbulence (Fig. 6.10):

$$\mathrm{Nu}_f = 0.26\,\mathrm{Re}_f^{0.6}\,\mathrm{Pr}_f^{0.34}\,Tu^{0.17}\,(\mathrm{Pr}_f/\mathrm{Pr}_w)^{0.25} \tag{6.5}$$

Note that the exponent on Tu in this relationship is 0.17.

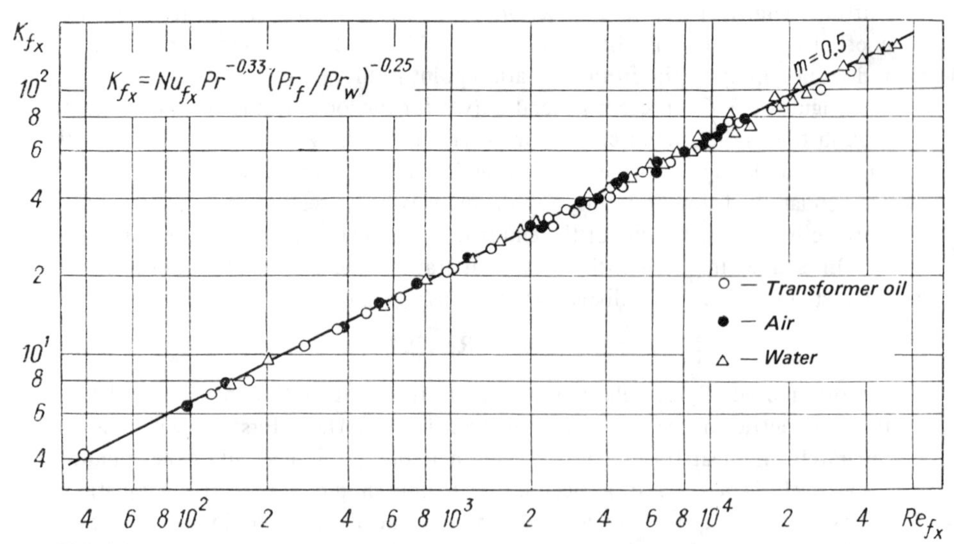

FIG. 6.9 Local heat transfer in the front part of a cylinder.

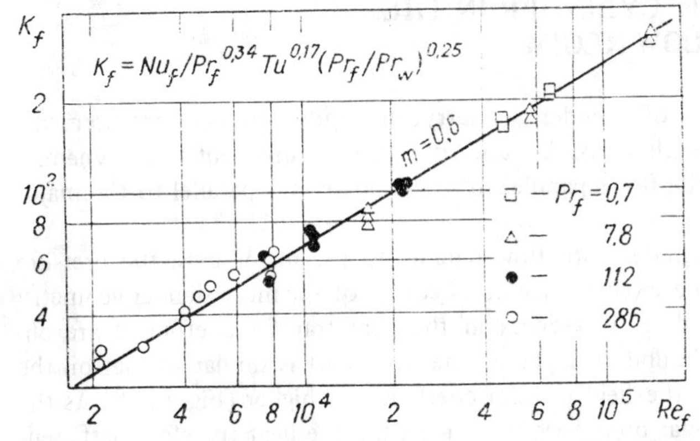

FIG 6.10 Average heat transfer on the front part of a cylinder for crossflow of various fluids at various turbulence levels.

The compilation of a similar relation for the rear part involves the difficulty of choosing the reference velocity value. We can only suggest a tentative equation for the average heat transfer on the rear part (Fig. 6.11):

$$\mathrm{Nu}_f = 0.068\,\mathrm{Re}_f^{0.73}\,\mathrm{Pr}_f^{0.4}\,(\mathrm{Pr}_f/\mathrm{Pr}_w)^{0.25} \tag{6.6}$$

Re and *Tu* in Eqs. (6.5) and (6.6) refer to the velocity at the minimum cross section of the channel and the characteristic length in the cylinder diameter. Our analysis of the data suggested that the influence of free-stream turbulence would occur only in the laminar part of the boundary layer.

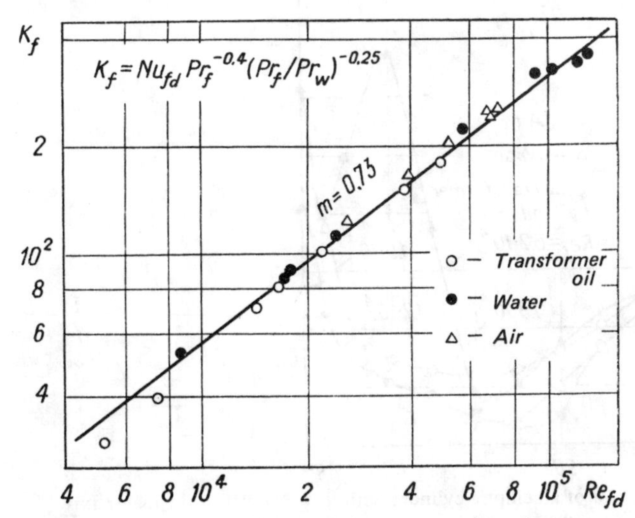

FIG. 6.11 Average heat transfer in the rear part of a cylinder.

6.2 THE ELLIPTIC CYLINDER IN THE SUBCRITICAL FLOW REGIME

To evaluate the effect of cylinder geometry, the study included measurements on elliptic cylinders with an aspect ratio of 1:2. Very different fluid dynamics were observed with the flows parallel to the minor versus parallel to the major axis.

On an elliptic cylinder with flow parallel to the major axis, the pressure drag is lower than on a circular cylinder, because of the more slender geometry (Chap. 4). Corresponding differences in the heat transfer coefficient are observed. The laminar boundary layer on the front part is similar to that on the circular cylinder, but the heat transfer coefficient is higher (Fig. 6.12). As the thickness of the laminar boundary layer increases, the heat transfer coefficient gradually decreases; the coefficient increases again after the separation at $\varphi \approx 100^\circ$. This separation occurs at a higher angular distance than for a circular cylinder. The critical flow regime is established at $Re_f = 6.2 \cdot 10^4$, and the heat transfer in the rear part of the elliptic cylinder increases as the flow becomes progressively more vortical. Heat transfer coefficients were 40–50% lower on the rear part than on the front part, for all the subcritical flows studied. A similar difference is possible on a circular cylinder only at very low Re_f.

Our studies of heat transfer with the various fluids showed that the elliptical cylinder gave no special effects regarding physical properties; thus, the effect of the fluid physical properties was described by introducing Pr_f to a power of 0.33. The free-stream velocity effect was described by Re_f to the power 0.5 and

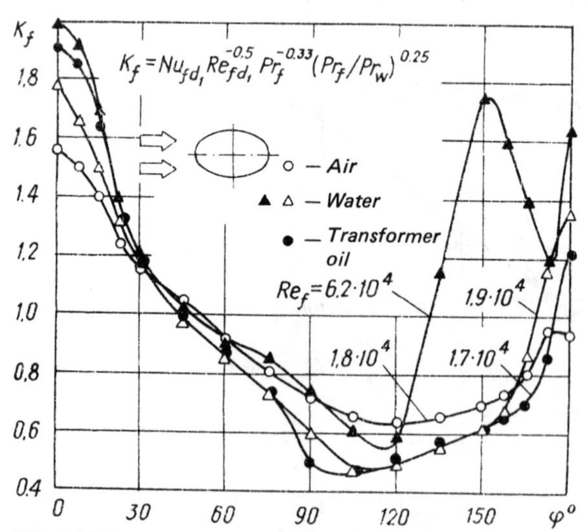

FIG. 6.12 Local heat transfer of an elliptic cylinder with flow parallel to the major axis for various values of Re_f and with $Tu \approx 1\%$.

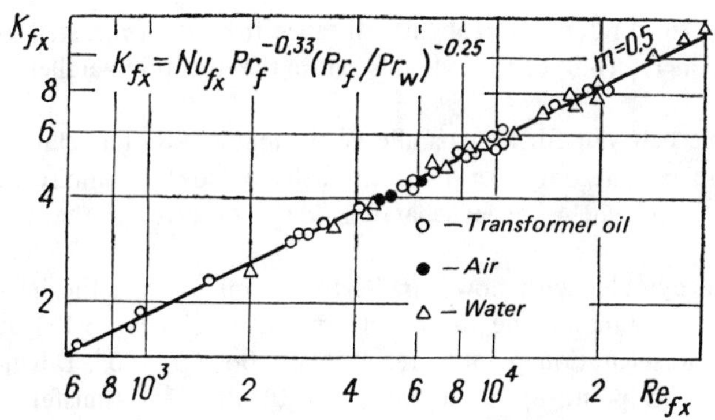

FIG. 6.13 A general curve for local heat transfer coefficient for an elliptic cylinder with flow parallel to the minor axis.

0.73 for the front and the rear parts, respectively. The exponents on Pr_f and Re_f are the same as for the circular cylinder.

For local heat transfer coefficient we suggest the expression

$$Nu_{fx} = 0.58\, Re_{fx}^{0.5}\, Pr_f^{0.33}\, (Pr_f/Pr_w)^{0.25} \tag{6.7}$$

for the front part (Fig. 6.13), and

$$Nu_f = 0{,}063\, Re_f^{0.73}\, Pr_f^{0.33}\, (Pr_f/Pr_w)^{0.25} \tag{6.8}$$

for the average heat transfer in the rear part (Fig. 6.14).

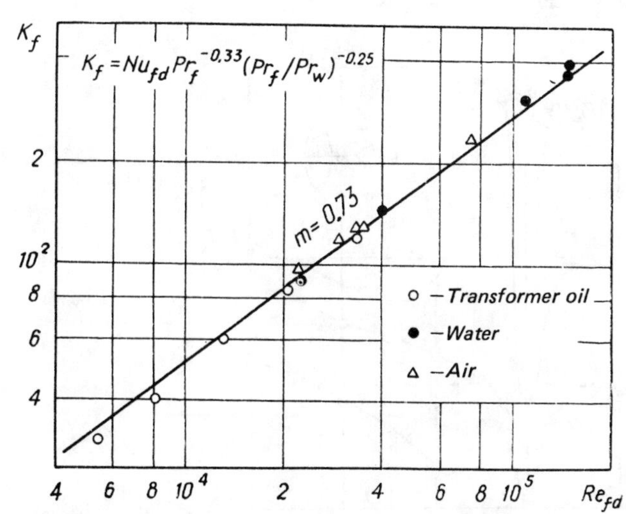

FIG. 6.14 Average heat transfer coefficient in the rear part of an elliptic cylinder with flow parallel to the major axis.

The values in Eq. (6.7) refer to the angular distance from the front stagnation point, and those in Eq. (6.8) to the cylinder axis in the direction parallel to the flow.

The fluid dynamic behavior for flow parallel to the minor axis of an elliptic cylinder is more complex. The flow over the front of the cylinder is reminiscent of an impact flow on a screen. A vortex separated flow is observed in the rear part.

For the elliptical cylinder with flow parallel to the minor axis, the heat transfer coefficient is constant on the front part up to $\varphi = 60°$ (Fig. 6.15). It then increases until the separation point is reached, at about $\varphi = 90°$. Downstream of the separation point up to about $\varphi = 120°$ the heat transfer is governed by the intensity of vortex shedding and, later, by the diffusing vortices and local boundary layers, which begin at the rear critical point. The analysis resulted in a heat transfer relationship similar to Eq. (1.1); values for the indices in this equation are presented in Table 6.3.

Note (in Table 6.3) the higher value of n for the rear part of the elliptic cylinder. Its value is similar to that for turbulent flow, which is quite reasonable for this geometry. All the studies of elliptic cylinders were performed with free-stream turbulence about 1%.

6.3 THE CIRCULAR CYLINDER IN THE CRITICAL FLOW REGIME

Our study of local heat transfer on circular cylinders included the critical flow regime for both air and water.

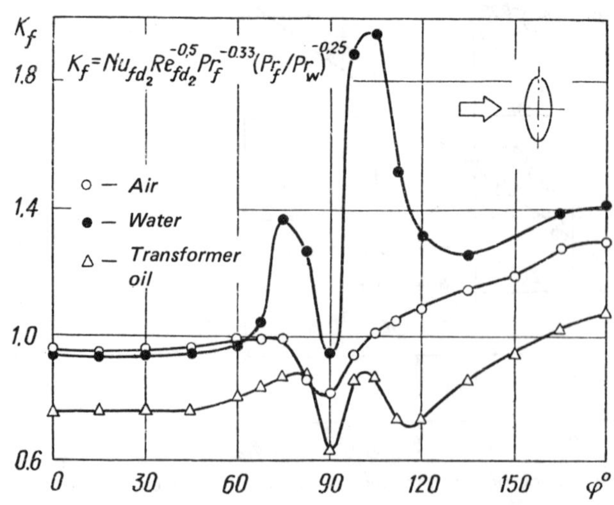

FIG. 6.15 Local heat transfer coefficient on an elliptic cylinder with flow parallel to the minor axis.

TABLE 6.3 Power Indices for Eq. (1.1) Describing Heat Transfer on an Elliptic Cylinder with the Flow Parallel to the Minor Axis

φ	c	m	n
0–90°	0.73	0.5	0.33
90–180°	0.086	0.73	0.4

Earlier critical region results by Kruzhilin [15] and by Schmidt and Wenner [16] for the lower critical ranges of Re showed that the laminar–turbulent transition occurred in the boundary layer at $\varphi \approx 85$–$90°$. A turbulent thermal boundary layer was also formed at this transition. The separation point was at $\varphi \approx 140°$, and the curves of the local heat transfer coefficient had two minima. In subsequent studies [24] over a wide range of critical and the supercritical flows, we demonstrated the existence of a relationship between the laminar–turbulent transition and the beginning of the turbulent thermal boundary layer on the one hand, and the Reynolds number and the turbulence level on the other (Figs. 6.16 and 6.18). Increasing both Re and *Tu* caused an upstream shift of the beginning of the turbulent thermal boundary layer to $\varphi = 30$–$35°$, so that it occupied a larger region on the front part of the cylinder. This shift was accompanied by a considerable augmentation of the heat transfer.

The curves in Figs. 6.16–6.19 illustrate the complex behavior of local heat transfer in the critical flow regime. The laminar boundary layer persists on the front part, and the heat transfer coefficient decreases downstream as the boundary-layer thickness increases, until the first minimum is reached.

However, the curves for the local heat transfer coefficient show some important differences when compared with those for subcritical flows (see Section 6.1). The curves for water and air exhibit two minima, which reflect the characteristic features of the dynamics of the flow over the surface in the critical flow regime. The importance of the dynamic behavior of the laminar–turbulent

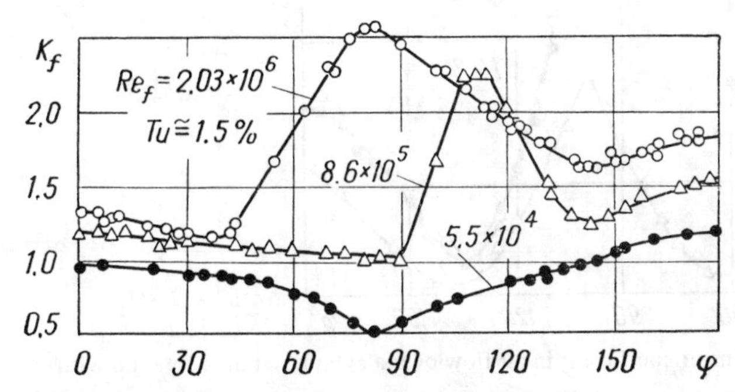

FIG. 6.16 Local heat transfer coefficient in water flow over a circular cylinder for $Tu = 1.5\%$ and variable Re_f.

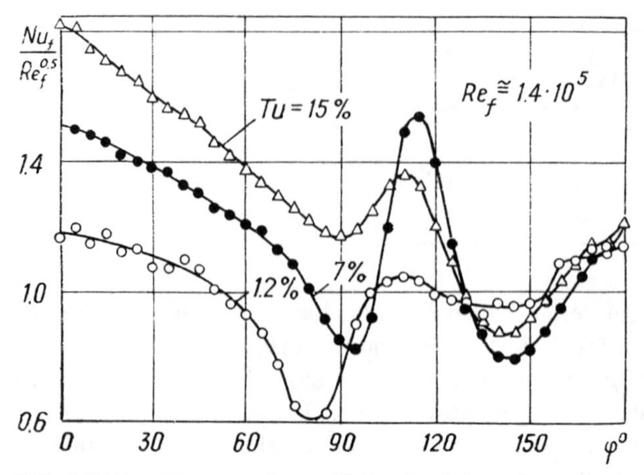

FIG. 6.17 Local heat transfer coefficient in air flow over a circular cylinder at various levels of turbulence.

transition in the boundary layer was stressed in Chap. 4. Juding from the available data, the beginning of the transitional region cannot be located unambiguously. Contradictory interpretations have been suggested by different authors. Some authors (e.g., Giedt [31]) maintain that, in the critical range of Re_f, the laminar boundary layer loses its stability and contains a transitional and

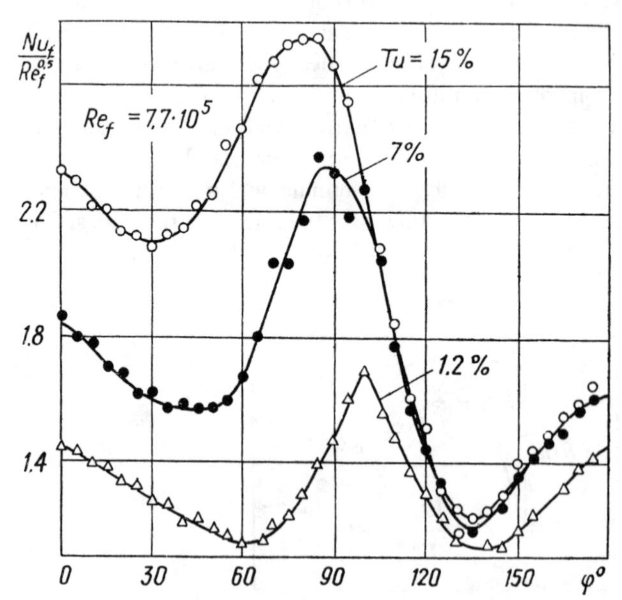

FIG. 6.18 Local heat transfer coefficient in air flow over a cylinder at high Re_f and with various turbulence levels.

a turbulent region, which are followed, above a certain value of Re_f, by boundary-layer separation. These authors interpret the two minima in the curves of the local heat transfer coefficient (Figs. 6.16-6.19) as the beginning of the transitional region and the separation of the turbulent boundary layer, respectively. Other authors (e.g., Roshko [27] and Tani [28]) suggest that the laminar boundary layer separates at $\varphi \approx 85$-$90°$ (i.e., at a similar position to that in subcritical flow). The boundary layer subsequently reattaches to form a separation bubble, and the subsequent turbulent boundary layer separates at $\varphi = 140°$. Here, the two minima are interpreted as being related to the two separation points; the first occurs, it is suggested, in the transitional region inside the separation bubble at $\varphi \approx 100°$ and the second occurs at the separation of the turbulent boundary layer at $\varphi = 140°$ (Fig. 6.17). At very high values of Re_f, the separation bubble is eliminated, and the first interpretation applies anyway. Thus, the two interpretations actually coincide for the critical flow regime.

Although we are supporters of the first interpretation [24], we would point out that there is unlikely to be much quantitative difference between the two interpretations, since the partial separation point and the transition point are very close together, occupying an interval of only 5-10° on the cylinder. The alternative interpretations of the fluid dynamic and thermal processes may, moreover, be useful for the development of a physical model and of the associated mathematical simulations.

With the onset of the critical flow regime, the separation bubble appears

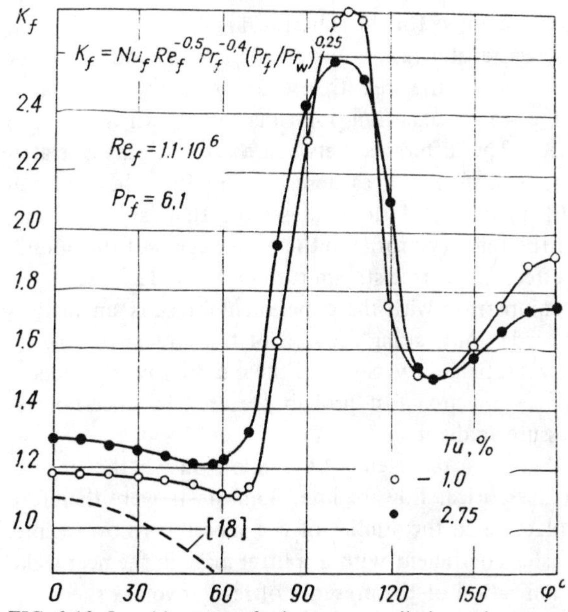

FIG. 6.19 Local heat transfer between a cylinder and water at various turbulence levels.

immediately after the separation of the laminar boundary layer. This phenomenon is characterized by a downstream shift of the laminar boundary layer separation (from $\varphi \approx 80\text{–}100°$, Fig. 6.17). The region of the surface that is covered by the laminar boundary is thus enlarged; the turbulent boundary layer separates from the surface at $\varphi \approx 140°$. When the critical flow regime is established at a low level of turbulence ($Tu \approx 1\%$), both the results of Achenbach [26] and those of our study suggest that the first minimum in Fig. 6.17 corresponds to the onset of the laminar-turbulent transition in the boundary layer inside the separation bubble, and that the second minimum is associated with the separation of the turbulent boundary layer.

With an increase of Re_f and/or Tu, the separation bubble diminishes in size and ultimately disappears, the transition point being shifted upstream and the laminar part of the boundary layer becoming shorter (Figs. 6.16, 6.18, and 6.19).

A general conclusion from our experiments is that turbulence, even at a very low level ($Tu \approx 1.0\%$), aids the onset of the critical flow regime. Local effects of the free-stream turbulence are also observed; the effects are most pronounced in the front stagnation point but persist for the whole front zone, decreasing with angular distance (Fig. 6.17 and 6.19). No systematic effects of free-stream turbulence on fluid dynamics are observed in the zone. The vortex flow region in the rear of the cylinder is obviously so highly turbulent as to be insensitive to external disturbances. The local heat transfer coefficient, which increases with the free-stream turbulence, is interpreted for the critical regime in a way similar to that described above for the subcritical regime.

A comparsion of the heat transfer results for air (Fig. 6.18) and for water (Fig. 6.19) shows that the effect of the free-stream turbulence is considerably greater in air, not only in the front stagnation point (as was shown in Chap. 5) but also on the whole surface. The difference between the two fluids is related to the fluid physical property differences, as described by Pr_f. The measurements of Kažimekas [116] in turbulent flows illustrated that, at higher Pr_f, there was a deformation of the velocity profile in the outer region of the boundary layer, thus causing the effect of the free-stream turbulence to decrease.

In order to provide a comparison with the experimental results, an analysis of local heat transfer was carried out taking account of free-stream turbulence (Fig. 6.20); the agreement was satisfactory. Some of the quantitative differences observed in the transitional region are most probably caused by uncertainties in the estimation of the pressure gradient.

At the present time, there is a divergence of opinion on whether regular vortex shedding occurs in the critical flow regime. Some tests were therefore performed with a splitter plate, as in the studies of the subcritical flow regime. A decrease of the heat transfer coefficient with a splitter plate in the near wake could be considered as an indication of the presence of regular vortex shedding. In fact (Fig. 6.21), in the critical flow regime for $\mathrm{Re}_f = 1.34 \cdot 10^5$, no changes in

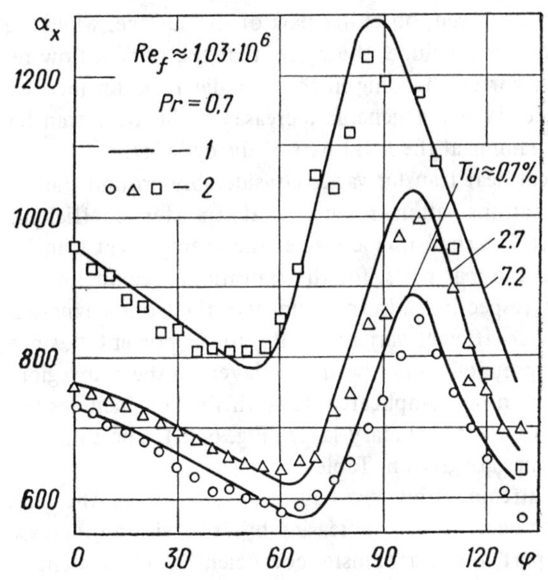

FIG. 6.20 Local heat transfer in the critical flow regime: 1, analysis; 2, experiment.

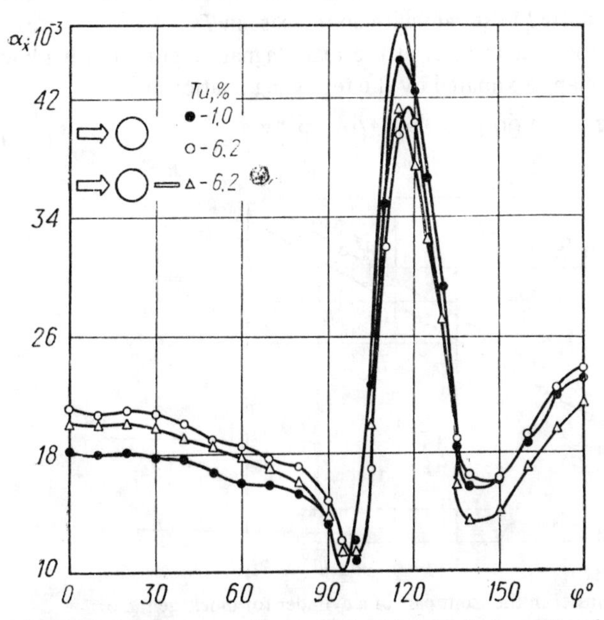

FIG. 6.21 Local heat transfer coefficient for a cylinder with and without the regular vortex shedding.

heat transfer coefficient were observed, on either part of the surface, when the splitter plate was introduced. We conclude, therefore, that the critical flow regime does not exhibit regular vortex shedding in the cylinder rear. An increase of the free-stream turbulence caused a general increase in the heat transfer coefficient, the maximum remaining at the front part of the cylinder.

As we have seen, the local heat transfer varies considerably around the circumference. Practical applications often require heat transfer coefficients averaged over various regions. We shall now consider the average heat transfer coefficients for the front and the rear parts, for the transitional region, and for the turbulent boundary layer, respectively. In the vicinity of the front stagnation (Fig. 6.22), the heat transfer coefficient varies with Re_f to the power 0.6, since this region is covered by the pseudo-laminar boundary layer. In the rear region. the fluid dynamic behavior is more complicated since this region includes the transitional region and the turbulent boundary layer (Fig. 6.23). The values of the power indices for these zones are given in Table 6.4.

An increase of the free-stream turbulence causes an increase of the heat transfer in the front part, the heat transfer increasing by 20% with an increase of *Tu* up to 7%. In the rear part, the heat transfer coefficient is independent of the free-stream turbulence, being dominated by the vortex flow (Fig. 6.23). At $\mathrm{Re}_f < 5 \cdot 10^5$, changes are observed in the relationships for heat transfer in the transitional and the turbulent regions. These may possibly be ascribed to the separation bubble. The region in which these changes occur is related to the turbulence level, and is shifted upstream with an increase in *Tu*.

Our results [117] for heat transfer at the rear stagnation point (with a low level of turbulence) were approximated by the following relationship:

$$\mathrm{Nu}_f = 0.09\,\mathrm{Re}_f^{0.83}\,\mathrm{Pr}_f^{0.4}\,(\mathrm{Pr}_f/\mathrm{Pr}_w)^{0.25} \tag{6.9}$$

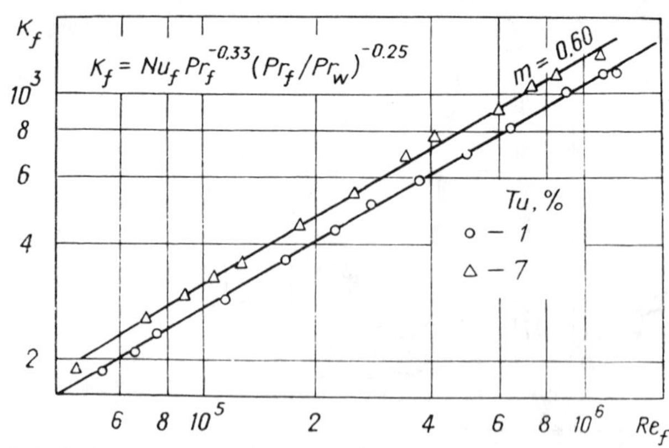

FIG. 6.22 Average heat transfer in the front part of a cylinder for blockage factor $k_q = 0.3$.

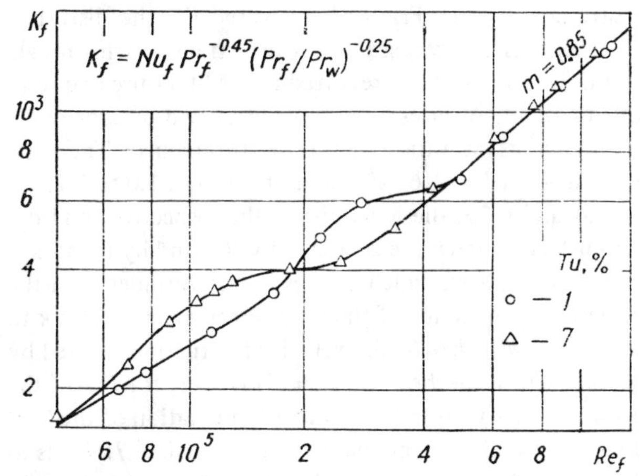

FIG. 6.23 Average heat transfer coefficient in the transitional and the turbulent regions for flow over a cylinder.

6.4 LOCAL HEAT TRANSFER BEHAVIOR FOR FLOW OVER A ROUGH-SURFACE CYLINDER

Heat transfer between a cylinder and a crossflow can be augmented by disturbing the boundary layer, which is the main source of the thermal resistance. The disturbance may be introduced either externally (e.g., by increasing free-stream turbulence) or by causing an increase of the turbulence in the boundary layer itself (e.g., by roughening the heat transfer surface). The presence of roughness on the heat transfer surface may even lead to a complete destruction of the boundary layer. Studies on the heat transfer and fluid dynamics in flow over rough-surface cylinders have been proceeding for many years, but the early studies were of the subcritical flows of air. Only comparatively recently (Achenbach [42]) has systematic data for the critical flow regime become available. Achenbach identified optimal geometries of the surface elements, which led to a 60–80% augmentation of the heat transfer, with only a 10% increase in the friction drag. Investigations of heat transfer in longitudinal flow over rough

TABLE 6.4 Constants and Power Indices for Average Heat Transfer in the Different Regions of a Cylinder

Region	c	m	n
φ_0–φ_p	0.32	0.6	0.33
φ_p–φ_s	0.03	0.9	0.45
φ_s–$\varphi_{180°}$	0.02	0.85	0.45

plates had indicated a strong effect of Pr_f; with increased Pr_f the disturbed boundary layer becomes thinner. However, studies of heat transfer with rough-surface cylinders in liquid flows had not been reported in the literature and were therefore included in our program. A series of tests were carried out for water flow over rough-surface cylinders at various levels of turbulence. The tests covered the range of Re_f from $4 \cdot 10^4$ to $1.6 \cdot 10^6$, and of Tu from 1.0 to 7%.

As shown in Figs. 6.24 and 6.25, the locations of the respective fluid dynamic regions on a rough-surface cylinder are governed by Re_f and by roughness height. The effect of surface roughness, which creates higher turbulence in the boundary layer, may be described in terms of the ratio of roughness height k to the boundary layer thickness δ. With $k \ll \delta$, the velocity fluctuations caused by the surface elements do not affect the heat transfer. However, for $k \geqslant \delta$, the velocity fluctuations in the boundary layer lead to an augmentation of the heat transfer. As shown in Figs. 6.24 and 6.25, an increase of Re_f and/or Tu leads to an earlier laminar–turbulent transition in the boundary layer. The critical height of the surface elements required to cause the transition is given by

$$k^+ = \frac{u_* k}{\nu} = 15 \tag{6.10}$$

In our tests, k^+ reached a value of 20 and was higher than the value given by Eq. (6.10), being therefore sufficient for the transition. Using our measurements of shear stress (see Chap. 4) to evaluate u_*, the value of k^+ can be calculated and found to increase from 0 at the front stagnation point up to the value related to the shear stress at $\varphi = 60°$. This implies a relationship between the development of the boundary layer and the effect of surface roughness. The location of the laminar–turbulent transition, as triggered by surface elements of the critical height, depends also on Re_f and on Tu. Our measurements (Chap. 4) indicated low absolute values of the shear stress in the front zone of the cylinder, and this may be the explanation for the low effect of the surface roughness in this region.

A minimum in heat transfer coefficient occurs in the rear part of the cylinder at the separation point location. This occurs at $\varphi = 130°$, earlier than on a smooth-surface cylinder, because the boundary layer is made thicker by the addition of streams separating the surface elements. The heat transfer coefficient

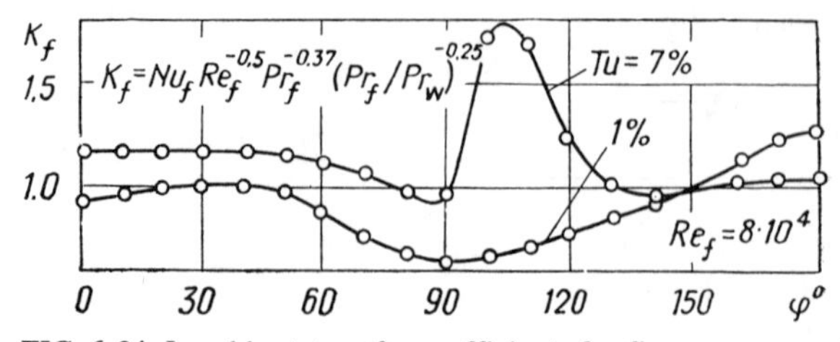

FIG. 6.24 Local heat transfer coefficients for flow over a rough-surface cylinder.

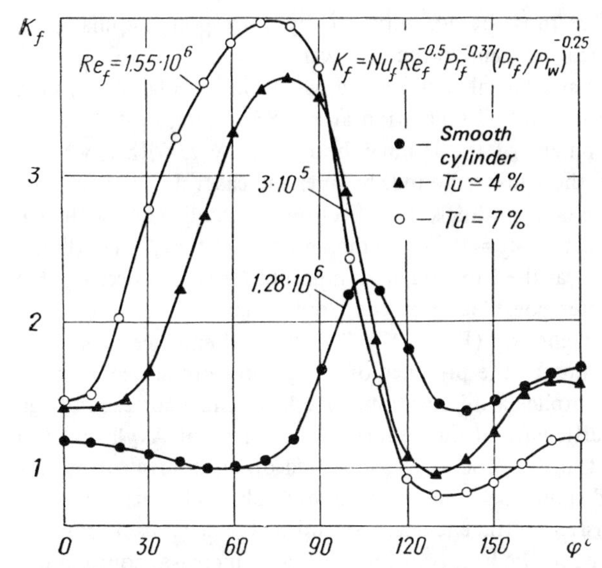

FIG. 6.25 Local heat transfer behavior of a rough-surface cylinder at varying turbulence level.

is lower in the rear part of a rough-surface cylinder (Fig. 6.25) than on a smooth-surface one, again because of the thicker boundary layer and the wider wake.

6.5 THE BLOCKAGE FACTOR AND LOCAL HEAT TRANSFER

Both in industrial heat exchangers and in most experimental rigs, the tubes are placed in channels. The effect of the channel walls is expressed by the blockage factor, which is the ratio of the cylinder diameter to the channel height, $k_q = d/H$. Our results on heat transfer presented in the preceding chapters all refer to small blockage factors k_q from 0.17 to 0.30, and mostly to $k_q = 0.25$. However, because of the importance of the blockage factor, a separate study was performed to evaluate its effect over a wider range. The blockage factor was varied in two ways—by varying the height of the channel for a constant cylinder diameter, and vice versa. The study covered Re_f from $5\cdot10^4$ to $4\cdot10^6$ and k_q from 0.28 to 0.7 and was carried out with water flow.

An increase of the blockage factor involves basic changes of the pressure distribution (Chap. 4) with consequential changes in the velocity distribution outside the boundary layer and in the velocity gradient. With an increase of k_q, the point of minimum pressure is shifted downstream to $\varphi = 110°$ and the minimum becomes deeper. This shift is accompanied by a sharp increase of the hydraulic drag (Fig. 4.35). Such significant changes in the fluid dynamic parameters on the surface must be accompanied by corresponding changes in the fluid velocity in the rear part and in the near wake.

Considering first the subcritical flow regime, we again see a laminar boundary layer on the front part, and its separation at $\varphi = 85°$. Here, heat transfer is governed by the development of the laminar boundary layer. With low values of k_q, the thickness of the laminar boundary layer increases downstream, and the heat transfer decreases accordingly (see Sections 6.1 and 6.3). With high values of the blockage factor ($k_q = 0.7$), the maximum heat transfer coefficient occurs at $\varphi \approx 50°$, and not at the front stagnation point as it is for lower k_q (Fig. 6.26). Higher heat transfer coefficients in the front region were also noted by Akylbaev et al. [39] at higher k_q (Fig. 6.27). This type of enhancement of the heat transfer process is due to the presence of a high pressure gradient and an accelerated flow, which implies a thinner boundary layer and a correspondingly higher heat transfer coefficient. In our studies, as in those of Akylbaev et al. [39], this type of heat transfer process was observed on an interval of $\varphi = 50$–$70°$. At higher angular distances, the boundary layer thickness begins to increase again, and the heat transfer coefficient decreases up to the separation point. In the rear part of the cylinder, the heat transfer coefficient increases continuously due to the action of the vortex flow; however, with moderate blockage factors (Fig. 6.27), secondary minima appear in the curves of local heat transfer coefficient against Re_m, indicating the approach of the critical flow regime. At high values of k_q, these minima are less clear and it is perhaps this fact that gave rise to the hypothesis by Akylbaev et al. [39] that it may be impossible to enter the critical flow regime with a high blockage factor. However, our studies, carried out on a wide range of Re_f, do not confirm this hypothesis.

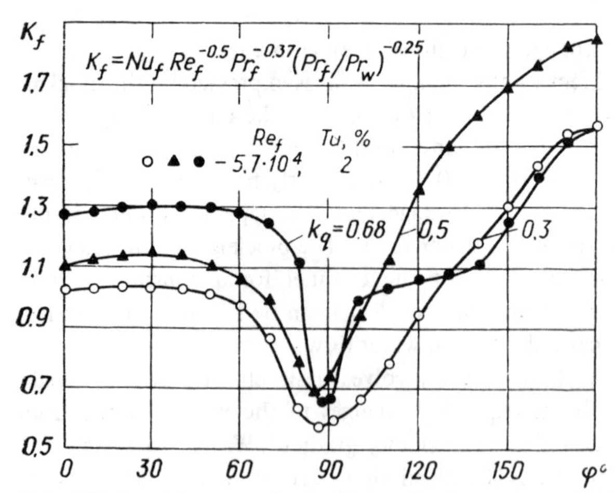

FIG. 6.26 Local heat transfer coefficient for flow over a cylinder in the subcritical flow regime and for a range of blockage factors.

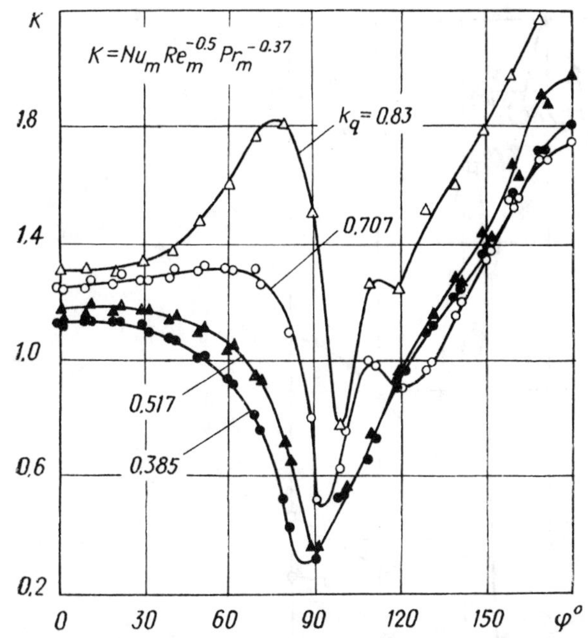

FIG. 6.27 Local heat transfer coefficient for flow over a cylinder at various values of the blockage factor [39].

The curves of the local heat transfer coefficient (Fig. 6.28) for flow over a cylinder in the critical regime [118], and with a blockage factor $k_q = 0.68$, show a shift of the position of the maximum coefficient to $\varphi = 60°$ for $Re_f = 5.84 \cdot 10^5$; with a further increase of Re_f, the maximum returns to the stagnation point. An increase of Re_f at constant diameter d leads to the increase in heat transfer coefficient at the front stagnation point, indicating more intensive flow in this zone. In the critical flow regime, two minima are observed in the curves of the local heat transfer coefficient. At certain values of Re_f inside the critical range, a separation bubble is also observed, due mainly to free-stream turbulence. The existence of this separation bubble causes the first minimum to shift upstream to $\varphi = 105°$. With the further increase of Re_f, the separation bubble disappears. Here, the laminar boundary layer on the front zone becomes unstable, so that, as was found in the results with low k_q, the transition point moves gradually upstream and eventually reaches $\varphi = 35°$. However, the laminar boundary layer persists on the front zone, although its thickness increases downstream for both low and moderate values of k_q (<0.5), and the heat transfer coefficient decreases accordingly. For high k_q and moderate Re_f (Fig. 6.28, curve for $Re_f = 5.84 \cdot 10^5$), because of the rapid acceleration of the flow with increasing φ, the maximum heat transfer coefficient is at $\varphi = 60°$, instead of at the front stagnation point.

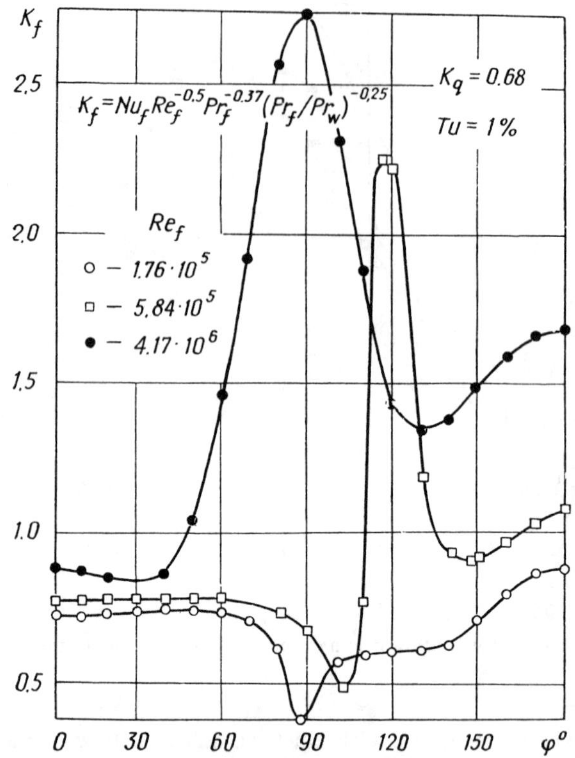

FIG. 6.28 Local heat transfer from a cylinder in a crossflow in the critical regime with a high blockage factor and at various Re_f values.

At still higher Re_f, when the separation bubble disappears, the velocity over the front zone becomes so high that the pseudo-laminar boundary layer loses its stability. In this case, notwithstanding the large k_q value, the laminar–turbulent transition is shifted upstream. The maximum heat transfer coefficient occurs here at $\varphi = 90°$. The turbulent flow zone in the rear is not affected by this further increase of Re_f, and the separation point of the turbulent boundary layer (the second heat transfer minimum) remains at $\varphi = 130$–$140°$.

Values of the local heat transfer coefficient (Fig. 6.29), measured at a constant $Re_f = 6.3 \cdot 10^5$ and constant H, show a considerable general increase at high k_q. The increase is most pronounced at $k_q > 0.5$. This behavior was observed also at some other values of Re_f. Increasing k_q at constant Re_f increases considerably the fraction of the surface covered by a turbulent boundary layer.

The above discussion reveals some significant differences in local heat transfer behavior for the subcritical and critical flow regimes. The major distinction lies in the different locations of the laminar–turbulent transition point (Fig.

6.28), both with and without the separation bubble. The exact location depends on Re_f, k_q, and Tu (Fig. 4.6). The effect of k_q is particularly pronounced at $Tu = 1\%$. With lower k_q, the laminar boundary layer is less stable, and the critical flow regime begins at lower Re_f. The stabilization at high k_q is due to the high pressure gradient. When k_q is increased at a given free-stream velocity, the velocity in the front zone rises abruptly; this is, of course, reflected in the behavior of the laminar-turbulent transition. However, at higher levels of turbulence ($Tu = 7\%$), the transition is less sensitive to k_q (Fig. 4.6), because the boundary layer becomes pseudo-laminar, under the influence of the turbulent fluctuations. The transition point is independent of the blockage factor.

The data on local heat transfer behavior illustrate well the contributions of the respective fluid dynamic regions on the cylinder. Since, in the subcritical flow regime, the thermal processes are well defined, both in the laminar boundary layer and in the rear, the average heat transfer is conveniently described by a binomial relationship:

$$Nu = c\, Re^{0.5} + c_1\, Re^{m_1} \tag{6.11}$$

This relation has been successfully applied by Perkins and Leppert [41] and by Akylbaev et al. [39] for the determination of the heat transfer coefficient in the respective regions. On the basis of experimental data, Akylbaev suggested alternative relationships for heat transfer coefficients for flow over a cylinder in the subcritical flow regime, namely,

$$Nu_m = 0.86\,(1 + 1.18\, k_q^3)\, Re_m^{0.5}\, Pr^{0.37} \tag{6.12}$$

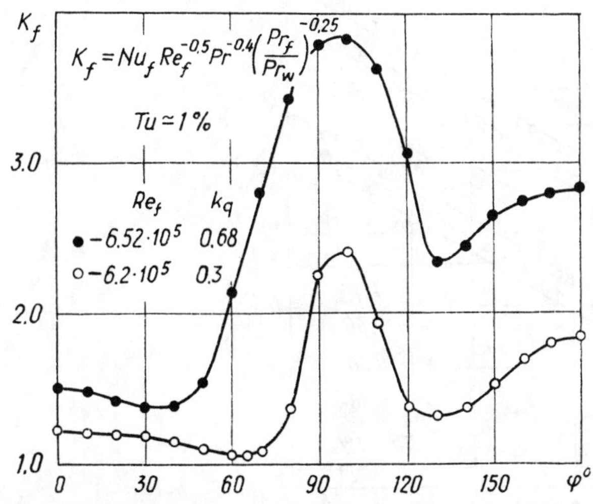

FIG. 6.29 Local heat transfer from a cylinder in the critical flow regime with various blockage factors.

for the front part and

$$\mathrm{Nu}_m = c(k_q)\,\mathrm{Re}_m^{0.7}\,\mathrm{Pr}_m^{0.37} \tag{6.13}$$

for the rear, where

$$c(k_q) = 0.120(1 + 1.18k_q^3)^{1.4} \quad \text{for} \quad k_q < 0.6$$

$$c(k_q) = 0.088(1 + 1.18k_q^3)^{1.4} \quad \text{for} \quad k_q > 0.6$$

Although Eqs. (6.12) and (6.13) are undoubtedly applicable for the subcritical flow regime, one cannot be sure of the validity of their extrapolation to the critical flow regime (Figs. 6.27 and 6.28) because of the very different nature of the processes in the two cases. In subcritical flow, the separation point at $\varphi \approx 90°$ lies between two symmetric regions of an equal size. In the critical flow regime, on the other hand, the presence of the two minima gives rise to three separate regions–laminar, turbulent and vortex–all of which depend in a complex way on Re, k_q, and Tu. It is more reasonable to determine the average heat transfer for each separate region by power-law relationships of the type

$$\mathrm{Nu}_f = c\,\mathrm{Re}^m\,\mathrm{Pr}^n\,Tu^k\,k_q^j\,(\mathrm{Pr}_f/\mathrm{Pr}_w)^{0.25} \tag{6.14}$$

The empirical data in Figs. 6.30–6.32 show that the main parameters in Eq. (6.14) vary significantly. The high level of turbulence is reflected by a higher

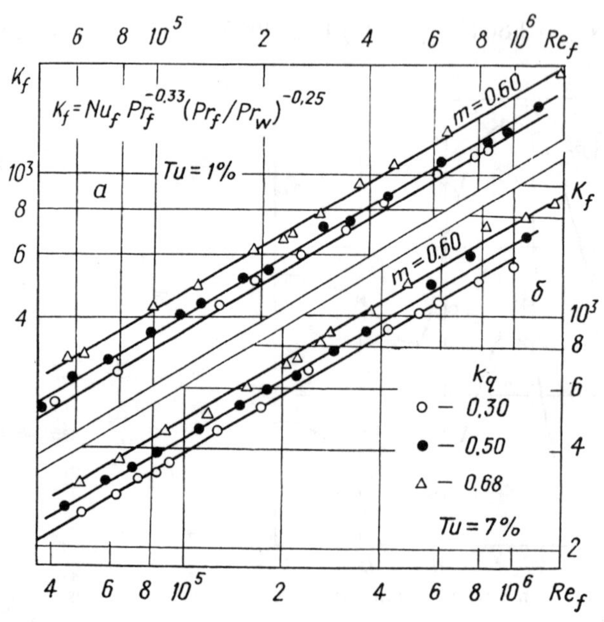

FIG. 6.30 Average heat transfer in the laminar region: α, $Tu = 1\%$; δ, $Tu = 7\%$.

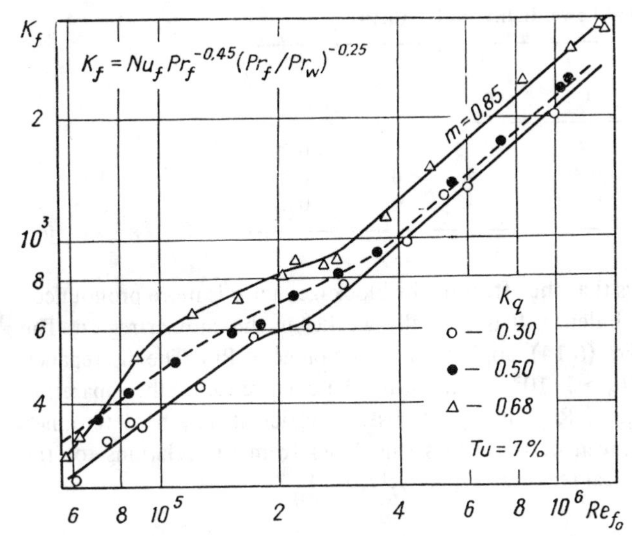

FIG. 6.31 Average heat transfer in the transitional and turbulent regions for flow over a cylinder with various blockage factors.

power index on Re_f and even for the laminar region, the power index m has a value (0.6) that is much higher than that in Eq. (6.12). Our experiments showed similar variations of the other parameters in Eq. (6.14) (see Table 6.5).

For the turbulent and the vortex regions (Figs. 6.31 and 6.32), the exponents on Re_f and Pr_f exceed the values ($m = 0.8$ and $n = 0.43$) characteristic of a turbulent boundary layer. This is probably due to the formation of the separation bubble and to the vortex processes in the wake, these phenomena being over the full range of blockage factor.

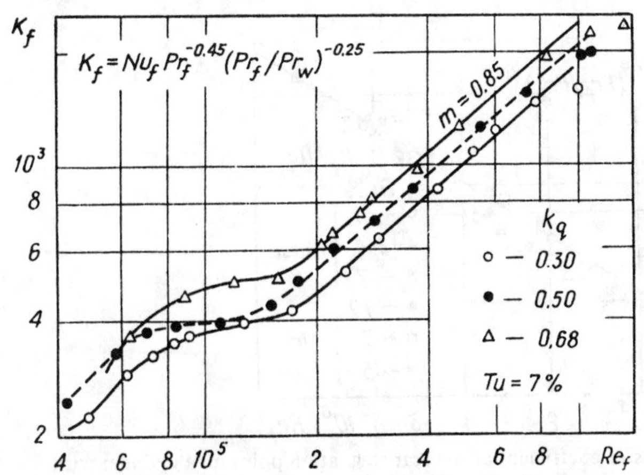

FIG. 6.32 Average heat transfer in the vortex region for flow over a cylinder with various blockage factors.

TABLE 6.5 Constants and Power Indices of Eq. (6.14)

Translate this head	c	m	n	k	j
S_0–φ_p	0.32	0.6	0.33	0.15	0.25
φ_p–φ_s	0.06	0.85	0.45	0.15	0.25
φ_s–$\varphi_{180°}$	0.05	0.85	0.45	0	0.25

Figure 6.30 shows that the effect of the blockage factor is more pronounced at lower levels of turbulence. It is generally weaker in the laminar region. The suggested equation, Eq. (6.14), applies for a wide range of Re_f. The discrepancy observed for $10^5 > Re > 3 \cdot 10^5$ is caused by the occurrence of the separation bubble. For this range of Re_f, the heat transfer coefficient may be determined from the arithmetic mean of the values calculated from the relations for the critical and the subcritical regimes, respectively.

6.6 HEAT TRANSFER AT THE REAR STAGNATION POINT

The rear stagnation point must be considered separately since the existing empirical and analytical equations for the heat transfer in this region differ greatly from those for other regions both in the determining parameters and in the absolute values calculated. The difference is sometimes as high as a factor of 2. On the basis of the experimental data, Richardson [119] suggested the following relationship for the heat transfer coefficient at the rear stagnation point:

$$Nu = 0.19\, Re^{0.66}\, Pr \tag{6.15}$$

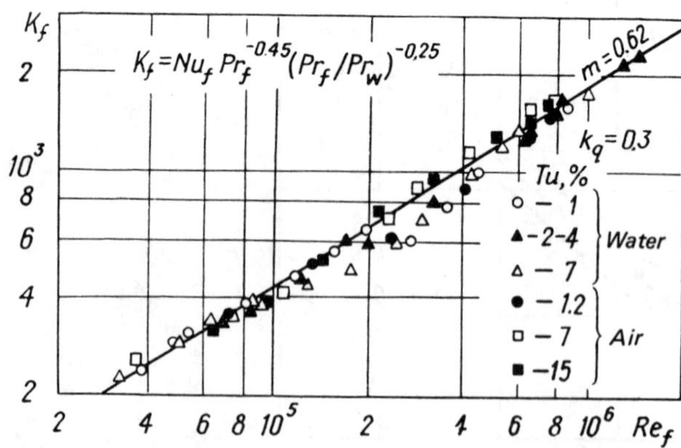

FIG. 6.33 Local heat transfer coefficient at the rear stagnation point on a cylinder with $k_q = 0.3$.

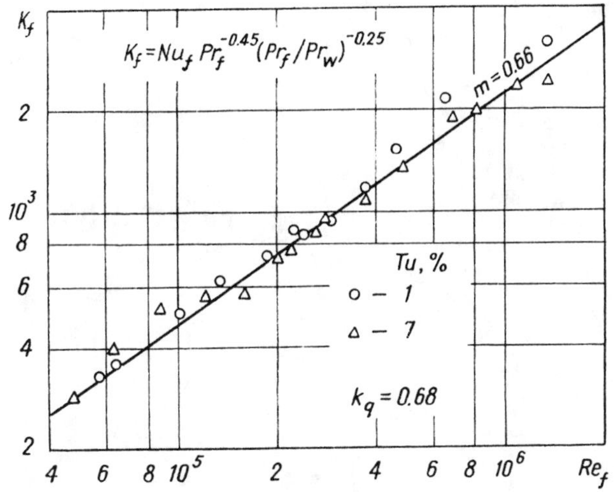

FIG. 6.34 Local heat transfer coefficient at the rear stagnation point for $k_q = 0.68$.

The following solution by Spalding [120] was based on the solution of the turbulent transfer equations:

$$Nu = 0.15\, Re^{0.6}\, Pr \tag{6.16}$$

The technique of Virk [44], employing the molecular transfer hypothesis, gave

$$Nu = 0.59\, Re^{0.5}\, Pr \tag{6.17}$$

Equations (6.16) and (6.17) underestimate the heat transfer coefficient. They suggest that the exponent on Pr is unity, whereas our experiments for the rear stagnation point indicate a value of 0.45. The value of m, the exponent on Re, is found to be 0.66 rather than 0.5. The two relations are only useful, therefore, as first approximations to the heat transfer behavior.

From the present unified experimental studies (which take account of the complex effects of free-stream turbulence) we suggest the following relationship for the heat transfer coefficient for $k_q = 0.68$:

$$Nu_f = 0.24\, Re_f^{0.66}\, Pr_f^{0.45}\, (Pr_f/Pr_w)^{0.25} \tag{6.18}$$

Comparisons in Figs. 6.33 and 6.34 show the rear stagnation point heat transfer coefficient increases at higher blockage factors. This is most probably related to changes in the parameters of the near wake, and primarily to changes in its width. The power index of Re_f also varies to a small extent. A further feature of the heat transfer at the rear stagnation point is its independence of the free-stream turbulence over the range $1 > Tu > 15\%$. The heat transfer at the rear stagnation point must therefore be governed by the vortex processes in the wake, rather than by the free-stream turbulence.

CHAPTER

SEVEN

AVERAGE HEAT TRANSFER

The average heat transfer coefficient for cylinders and other bodies in crossflow is governed by the free-stream velocity and turbulence, fluid physical properties, temperature difference, heat flux direction, diameter, and other factors. A general dimensionless equation for the average heat transfer coefficient is

$$\mathrm{Nu} = f\left(\mathrm{Re},\ \mathrm{Pr},\ \mathrm{Tu},\ \frac{\mu_f}{\mu_w},\ \frac{\lambda_f}{\lambda_w},\ \frac{c_f}{c_w},\ \frac{\rho_f}{\rho_w},\ \ldots\right) \tag{7.1}$$

Our studies of average heat transfer in different fluids included the effect of the above factors in the Prandtl and Reynolds number ranges $0.71 > \mathrm{Pr} > 320$ and $1 > \mathrm{Re} > 2 \cdot 10^6$.

Extensive graphical presentations of the experimental results are given here, and the results have been interpreted using a variety of functions. A large fraction of the data is presented in the Appendix.

7.1 THE AVERAGE HEAT TRANSFER AND THE FLUID PHYSICAL PROPERTIES

An interpretation of experimental results using Eq. (7.1) usually involves a general relationship:

$$\mathrm{Nu} = c\,\mathrm{Re}^m\,\mathrm{Pr}^n \tag{7.2}$$

The effect of the fluid physical properties on the heat transfer is expressed by the value of Pr, which is constant for gases of equal atomicity. Therefore the heat transfer relationship for air and other diatomic gases is reduced to

$$\mathrm{Nu} = f(\mathrm{Re}) \tag{7.3}$$

To illustrate the importance of the Prandtl number, we present in Fig. 7.1 our results for the average heat transfer coefficient between a 12-mm-diameter circular cylinder and a crossflow of various fluids [22]. The physical properties in the Nusselt number were referred to the bulk flow temperature. We note that the heat transfer coefficient is higher in transformer oil for wall-to-fluid heat transfer than for fluid-to-wall heat transfer: the experimental points for transformer oil are separated depending on the oil temperature. A similar situation was observed with water. The temperature and the type of fluid are the determining factors for the separation of the experimental points, and the variations may be expressed in terms of Prandtl number.

The results for both wall-to-fluid and fluid-to-wall heat transfer in air were compared to the results of other authors [3, 13]. The Prandtl number being constant in the two cases, no separation of the points was observed.

Thus any interpretation of the heat transfer must include the value of Pr_f with a corresponding exponent. The slopes of the curves in Fig. 7.1, which are 0.5 for $\mathrm{Re}_f < 10^3$ and 0.6 for $\mathrm{Re}_f > 10^3$, correspond to the exponent of Re_f in the interpretation equation.

The choice of n, the exponent of Pr_f, is equally important. In experiments with various viscous fluids, the Prandtl number varies widely, from 1 to

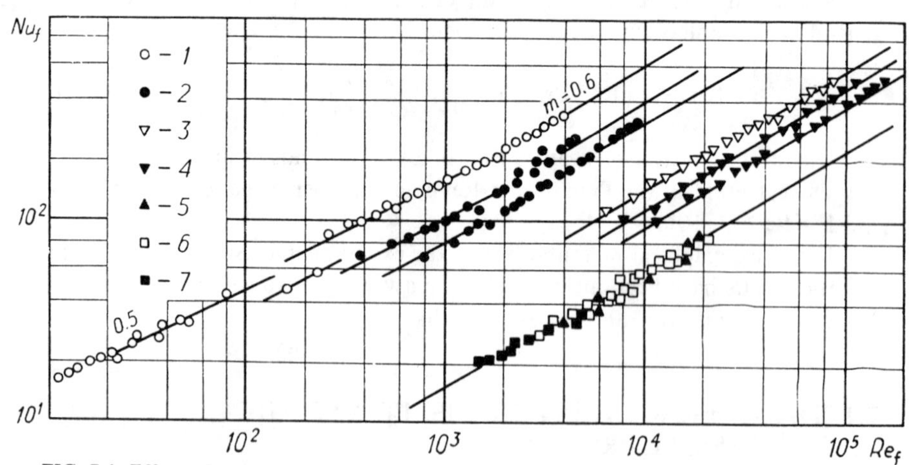

FIG. 7.1 Effect of physical properties on average heat transfer in crossflow over a cylinder: 1, 2, 5, wall-to-fluid heat transfer in flows of transformer oil, water, and air, respectively; 2, 4, 5, fluid-to-wall heat transfer in flows of the same; 6, 7, experimental data in the flow of air, from [3, 13].

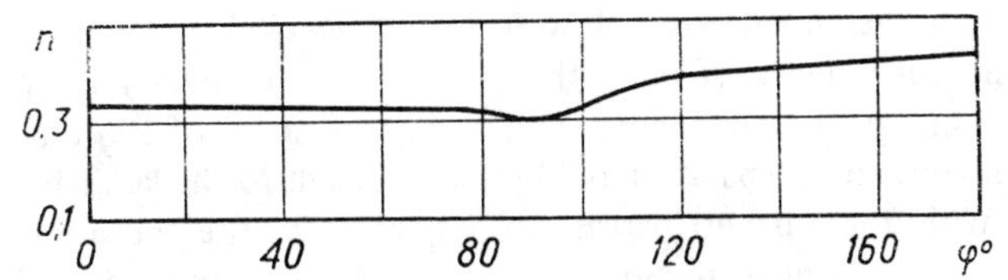

FIG. 7.2 Variation of the power index of Pr_f.

10^4, and an incorrect value for n may lead to considerable errors. Numerous publications suggest $n = 0.31$-0.33 for tubes in crossflow. This value was based on analytical solutions for laminar boundary layers on plates. Numerous experiments and analytical solutions [79, 121] have shown that n varies depending on the structure of the boundary layer; thus $n = 0.33$ and $n = 0.43$ for a turbulent boundary layer on a plate. A probable average value of n for a cylinder is the average between these two values.

In our studies of the local heat transfer from a cylinder [111] for constant wall heat flux q_w (Fig. 7.2) in different fluids, a variation of n around the circumference was also observed. For average heat transfer coefficients in the front and rear parts, $n = 0.33$ and $n = 0.4$, respectively.

Various studies of the laminar and the turbulent boundary layers have shown that n is relatively insensitive to Pr_f. For instance, the analytical study of heat transfer from a cylinder of Makarevičius [122] gave $n = 0.37$ for $Pr < 10$ and $n = 0.35$ for $Pr > 10$ at the front stagnation point.

The results shown in Fig. 7.1 for heat transfer between circular cylinders and various fluids were also interpreted for $10^3 > Re_f > 10^5$ by

$$Nu_f \cdot Re_f^{-0.6} = f(Pr_f) \tag{7.4}$$

as shown in Fig. 7.3. Here, the slope corresponds to the exponent of Pr_f and, in this range of Re_f, with constant t_w, the value of the exponent is about 0.37, and shows a tendency to decrease at the high Pr_f. Similar values of n were found for the lower range of Re_f.

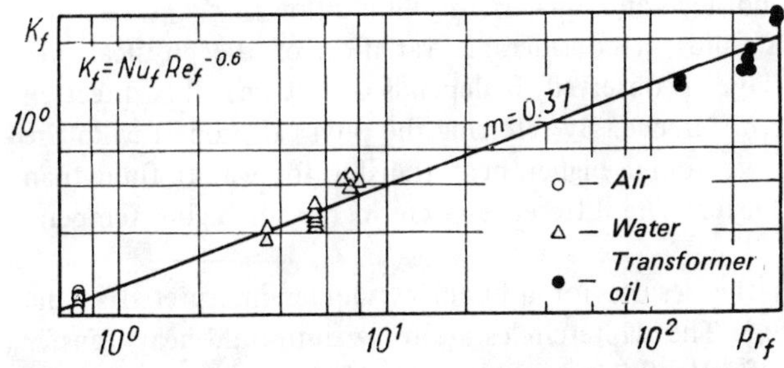

FIG. 7.3 Average heat-transfer coefficient as a function of Pr_f: ○, air; △, water; •, transformer oil.

The effect of Pr_f in the supercritical flow regime at $Re_f > 1.5 \cdot 10^5$ was studied separately. The dependence of n on the nature of the boundary layer was mentioned above. In the critical flow regime, at certain values of Re_f and Tu, the boundary layer contains a separation bubble, which leads to the laminar–turbulent transition. With the further increase of Re_f and Tu, the transition point moves upstream to the front stagnation point. Thus, since the value of n is higher for turbulent than for laminar flows, the effect of Pr_f must be more pronounced in the supercritical flow regime because of the more widespread turbulent boundary layer. Our experiments showed that the value of n was 0.4 for the supercritical flow regime, after the disappearance of the separation bubble, which contrasts with the value of 0.37 obtained [38] for the subcritical flow regime.

For the specific turbulent and the separation regions on the cylinder surface, $n = 0.45$ due to the high turbulence of the flow.

The process of heat transfer involves a variation of temperature and, consequently, variation of fluid physical properties. To properly evaluate the effect of the physical properties, it is necessary to evaluate the effect of the variation of the properties with temperature in the boundary layer or, in other words, to choose a correct reference temperature for the physical parameters. At present, two different approaches to the choice are equally popular. Numerous authors refer the physical properties of the fluids to the bulk flow temperature, and introduce a special parameter in the heat transfer relationships. Others refer the physical properties to the mean between the wall and bulk flow temperature, and employ these properties in equations developed for constant physical properties. Our comparison of equations for heat transfer coefficient developed by different authors for gas flows revealed that differences in the choice of the reference temperature constituted the main source of discrepancy.

In our heat transfer experiments with single tubes [13] and bundles of tubes [123] in gases, a sufficiently accurate interpretation of the heat transfer was achieved without any special term for physical properties, simply by referring the physical properties to the bulk flow temperature. We find the bulk flow temperature more convenient and sufficiently accurate for moderate fluid temperature (t_f) and for both gas and low-voscisity liquid flows.

In flows of viscous fluids, a considerable variation of the physical properties in the boundary layer is observed. It depends on the heat flux direction and on the temperature difference. By referring the physical properties to the bulk flow temperature, we found higher heat transfer for wall-to-fluid than for fluid-to-wall heat transfer. The differences were higher for higher temperature difference.

Figure 7.4 presents the results for a circular cylinder interpreted in the form $Nu_f/Pr_f^{-0.37} = f(Re_f)$. The white circles are for wall-to-fluid heat transfer and are higher than those for the fluid-to-wall heat transfer.

For sharp changes of the fluid physical properties in the boundary layer,

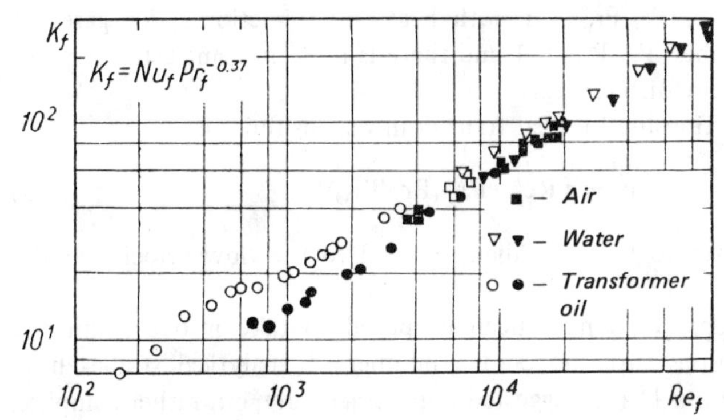

FIG. 7.4 Heat transfer between a cylinder and a crossflow of various fluids: ■, air, ▽▼, water, ○ ●, transformer oil.

Mikheev suggested [124] the use of the ratio Pr_f/Pr_w as a correlating parameter. It is sometimes replaced by the ratio μ_f/μ_w; for viscous fluids, $Pr_f/Pr_w \approx \mu_f/\mu_w$ as the property showing greatest variation is the viscosity. However, Pr_f/Pr_w is more reasonable, because the heat transfer is governed by the temperature field, and this is described by the Prandtl number. Thus, the value of Nu calculated from the bulk temperature (outside the boundary layer) is multiplied by Pr_f/Pr_w to an experimentally or analytically determined exponent p. Our experimental and analytical studies [121] gave p values of 0.25 and 0.19 for wall-to-fluid and fluid-to-wall heat transfer, respectively, in the laminar boundary layer on a plate.

For turbulent flows on a plate, p was equal to 0.25 and 0.17 [79] for wall-to-fluid and fluid-to-wall heat transfer, respectively.

Figure 7.5 illustrates the relationship between the heat transfer coefficient and Pr_f/Pr_w for the cylinder case. The slopes give $p = 0.25$ and $p = 0.20$ for wall-to-fluid and fluid-to-wall heat transfer, respectively. Judging by these experiments, the power index of Pr_f/Pr_w is higher for wall-to-fluid heat transfer, but for moderate temperature differences, a value of $p = 0.25$ may be assumed for the two cases.

If the physical properties in Re_f, Nu_f, and Pr_f are calculated at the bulk flow temperature, the effect of variable physical properties can be represented

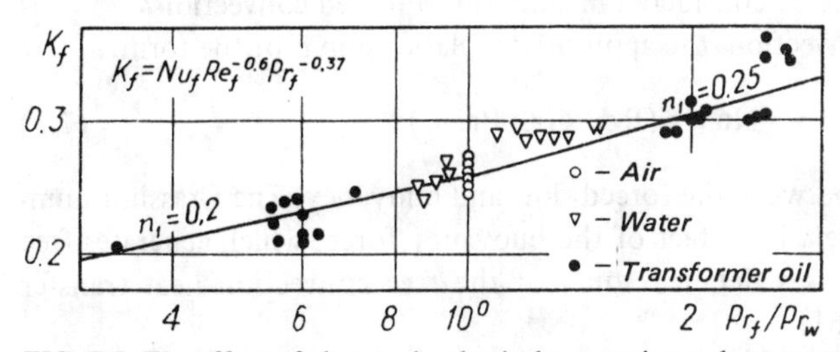

FIG. 7.5 The effect of changes in physical properties on heat transfer from a cylinder.

adequately in terms of Pr_f/Pr_w for both heat flux directions. For gases of constant Prandtl number, $Pr_f/Pr_w \approx 1$ and there is no significant effect, except at very high temperature differences.

The resulting relationship for heat transfer in viscous fluids is

$$Nu_f = c\, Re_f^m \cdot Pr_f^n (Pr_f/Pr_w)^p \tag{7.5}$$

where the values refer to the tube diameter and to the flow velocity at the minimum flow area.

Several other approaches have been proposed to take into account the fluid physical properties variation, and a number of analytical studies have been published. Gregorig [125] suggested a more accurate, but rather complex, empirical expression for the effect of fluid physical property variation in the boundary layer.

Both the constant c and the exponents in Eq. (7.5) depend on the flow regime. The exponent on Pr_f varies only from 0.37 to 0.4, but the exponent on Re_f varies more significantly, being 0.4 for the lower range of Re_f, and up to 0.8 and more for the supercritical flow regime [26]. Thus the relation between the Reynolds number and average heat transfer coefficient deserves special attention.

7.2 THE REYNOLDS NUMBER AND THE AVERAGE HEAT TRANSFER COEFFICIENT

Since the fluid dynamics of flow around a cylinder are highly dependent on Re, a corresponding sensitivity must be observed in the heat transfer.

The Low Range of Re

At low values of Re_f, heat transfer between a fluid and a cylinder is strongly influenced by free convection, which is overshadowed by forced convection at higher Reynolds numbers.

The question naturally arises: When can free convection be ignored, and when must its influence be considered in addition to forced convection?

For combined convection, the appropriate relationship is of the form:

$$Nu = f(Re,\ Gr,\ Pr,\ \varphi) \tag{7.6}$$

where φ is the angle between the forced flow and buoyancy. The Grasshof number Gr describes the relative effect of the buoyancy force, which generates free convection motion. For a detailed study of the free convection heat transfer see references [48, 126].

In combined convection, several alternative ways of accounting for free convection have been suggested. A number of authors [126, 127] have studied combined convection for flows over cylinders with $\varphi = 90$ or $180°$. Hatton et al. [127] introduced the concept of an effective Reynolds number Re_f related to φ by

$$Re_{ef}^2 = Re^2 \left[1 + 2.06 \frac{(Gr \cdot Pr)^{0.418}}{Re} \cos\varphi + 1.06 \frac{(Gr \cdot Pr)^{0.836}}{Re^2}\right] \quad (7.7)$$

and suggested the following relationship for the combined convection heat transfer:

$$Nu\,[T_m/T_f]^{-0.154} = 0.384 + 0.581 \cdot Re_{ef}^{0.439} \quad (7.8)$$

It is recommended that free convection be taken into account at $\varphi = 90°$ for $Re < 2.2(Gr \cdot Pr)^{0.418}$, and at $\varphi = 0$ and $180°$ for $Re < 10(Gr \cdot Pr)^{0.418}$.

As Re is increased, the influence of free convection becomes insignificant, and Eq. (7.8) reduces to

$$Nu\,[T_m/T_f]^{-0.154} = 0.384 + 0.581\, Re^{0.439} \quad (7.9)$$

An alternative approach is to use the technique of vectorial summation [128], the average heat transfer coefficient being found as the sum of the free convection heat transfer and the forced convection heat transfer coefficients, determined separately, the sum being given by

$$(Nu - 0.35)^2 = (0.24\, Gr^{1/8} + 0.41\, Gr^{1/4})^2 + + (0.5\, Re^{0.5} + 0.001\, Re)^2 \quad (7.10)$$

which reduces to

$$(Nu - 0.35)\sqrt{1 - \left[\frac{0.24\, Gr^{1/2} + 0.41\, Gr^{1/4}}{Nu - 0.35}\right]^2} = 0.5\, Re^{0.5} \quad (7.11)$$

Thus, Eq. (7.11) suggests, that with increasing Re, the heat transfer Nusselt number asymptotically approaches its forced convection value, and can be defined by a simplified two-term relationship:

$$Nu = 0.35 + 0.5\, Re^{0.5} \quad (7.12)$$

Collis and Williams [8] measured the heat transfer from wires in crossflow of air at low Re, and suggested a two-term heat transfer equation for the range $0.02 < Re < 44$:

$$Nu = 0.24 + 0.56\, Re^{0.45} \quad (7.13)$$

and suggested a single-term equation for the range $44 < Re < 140$:

$$Nu = 0.48\, Re^{0.51} \quad (7.14)$$

The results of Hilpert [12] for wires and tubes in crossflow of air are described for $1 < Re < 4$ by

$$Nu = 0.875\, Re^{0.31} \quad (7.15)$$

and for $4 < Re < 40$ by

$$Nu = 0.785\, Re^{0.39} \tag{7.16}$$

These relations for heat transfer in air agree to within 5% for the lower range of Re. More significant differences are observed in the exponent m on Re. A graphical representation of the results by different authors, reduced to the form $Nu = f(Re)$, showed that for $Re < 40$, $m = 0.40$, and that for $Re > 40$, $m = 0.5$. In liquid flows the trends were similar.

As a result of development of hot wire and other measuring devices for turbulent fluctuations, many recent publications deal with heat transfer from the wires which form the working elements in such devices. We measured average heat transfer coefficients for wires and cylinders in crossflow of transformer oil at low Re, and described our results by the expression

$$K_f = Nu_f \cdot Pr_f^{-0.37} \cdot (Pr_f/Pr_w)^{-0.25} = f(Re_f) \tag{7.17}$$

These results are compared in Fig. 7.6 with those of Davis [5]. Good agreement is observed, and the exponents on Re are again found to be 0.4 for $Re < 40$ and 0.5 for $Re > 40$.

The relationships recommended for calculations in this region are

$$Nu_f = 0.76\, Re_f^{0.4} \cdot Pr_f^{0.37} \cdot (Pr_f/Pr_w)^{0.25} \tag{7.18}$$

for $Re < 40$ and

$$Nu_f = 0.52\, Re_f^{0.5} \cdot Pr_f^{0.37}\, (Pr_f/Pr_w)^{0.25} \tag{7.19}$$

for $40 < Re < 10^3$.

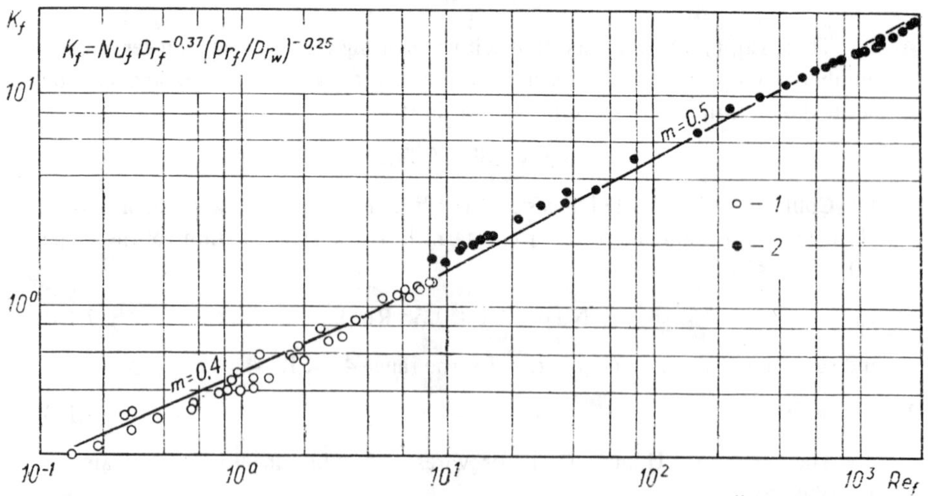

FIG. 7.6 Heat transfer of cylinders in the low range of Re_f: 1, Davis [5]; 2, Žukauskas [7].

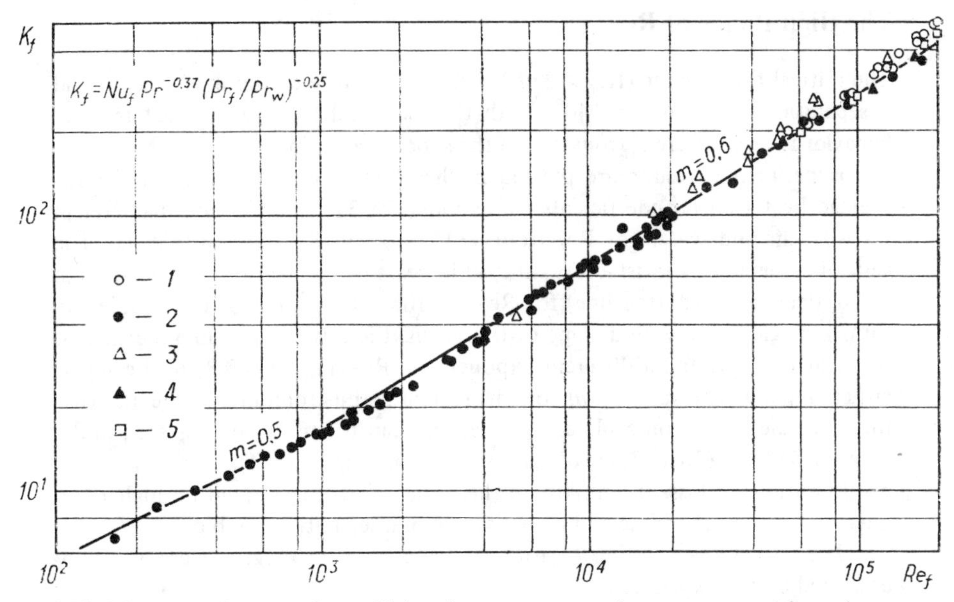

FIG. 7.7 Average heat transfer coefficient for constant t_w and constant q_w and for various fluids: 1, [24]; 2, [23]; 3, [111]; 4, [38]; 5, [25].

For low values of Re, Eq. (7.18) can be replaced by

$$\mathrm{Nu}_f = (0.35 + 0.62\,\mathrm{Re}_f^{0.4})\,\mathrm{Pr}_f^{0.37}\,(\mathrm{Pr}_f/\mathrm{Pr}_w)^{0.25} \tag{7.20}$$

The Subcritical Flow Regime

This regime features a laminar boundary layer on the front part of a cylinder. The boundary layer separates at $\varphi = 82°$, and gives rise to vortex shedding and a vortex flow. This characteristic fluid dynamic structure persists up to the critical range ($\mathrm{Re}_f = 2 \cdot 10^5$). The key characteristics of the average heat transfer must also be preserved throughout the subcritical flow regime.

Figure 7.7 presents the data in the form of Eq. (7.17) and includes results for the average heat transfer coefficient for a tube in air, water, and transformer oil, for constant t_w (the results shown also in Fig. 7.1), and constant q_w, and for both fluid-to-wall and wall-to-fluid heat transfer. The results of the various experiments are in good agreement and fall on a uniform curve. The average heat transfer coefficient for constant q_w [calculated using Eq. (2.3)] is in satisfactory agreement with the results for constant t_w. The slope of the line in Fig. 7.7 changes from 0.5 to 0.6 at $\mathrm{Re}_f = 10^3$, and the recommended relationship for heat transfer $10^3 < \mathrm{Re} < 2 \cdot 10^5$ is

$$\mathrm{Nu}_f = 0.26\,\mathrm{Re}_f^{0.6}\,\mathrm{Pr}_f^{0.37}\,(\mathrm{Pr}_f/\mathrm{Pr}_w)^{0.25} \tag{7.21}$$

The High Range of Re

The critical flow regime ($Re_f > 2 \cdot 10^5$) involves a turbulent boundary layer, and a separation point at $\varphi = 140°$. Both the local and the average heat transfer behavior is very different from that in the subcritical region.

Experimental results are plotted in the form of Eq. (7.17) in Fig. 7.8 for average heat transfer coefficients for cylinders of 31.1 and 50 mm diameter in flows of air and water for Re_f from $4 \cdot 10^4$ to $2 \cdot 10^6$, and constant q_w. The general agreement is satisfactory, and an increase of the heat transfer is noted in the critical range (dotted line) for $Re \approx 2 \cdot 10^5$. The points for $Re_f > 3.5 \cdot 10^5$ form a single curve with a slope of 0.8. Thus, there is considerable increase of the heat transfer, and a different exponent on Re_f (i.e., $m = 0.8$) in the supercritical regime. The value of m for this regime is thus found to be equal to that for a turbulent flow on a plate. A deviation from the main curve is seen in the range $1.5 \cdot 10^5 > Re > 3.5 \cdot 10^5$ is also seen in Fig. 7.8. This deviation is a manifestation of loss of stability in the critical flow regime, with a higher frequency of vortex shedding and the existence of a separation bubble.

The results given in Fig. 7.8 for the supercritical flow regime were originally described by the expression

$$Nu_f = 0{,}023\, Re_f^{0.8}\, Pr_f^{0.37}\, (Pr_f/Pr_w)^{0.25} \tag{7.22}$$

However, a more detailed analysis suggested that n, the exponent on Pr_f, should be 0.4 rather than 0.37.

7.3 SUMMARY OF CORRELATIONS FOR THE AVERAGE HEAT TRANSFER COEFFICIENT

The experimental data presented above were obtained with the turbulence level around 1%. This data is in good agreement with other results published, as shown in Fig. 7.9. For moderate temperature differences we suggest the following general relationship for average heat transfer coefficient:

$$Nu_f = c\, Re_f^m \cdot Pr_f^{0.37} \cdot (Pr_f/Pr_w)^{0.25} \tag{7.5}$$

where the values of c and m are given in Table 7.1. For gaseous flows the relation is simplified: e.g., for air, $Pr_f = 0.7$ = constant, and $Pr_f^{0.37} \cdot (Pr_f/Pr_w)^{0.25} = 0.88$.

For gaseous flows with $Re \approx 1$, the heat transfer can be calculated by the equations given in Section 7.2.

A number of alternative general relationships are given in the literature. In [41], two-form equations were suggested for each Re range. These describe the heat transfer in the front laminar boundary layer and that in the rear part of the cylinder, respectively. In a review of results of different authors, Churchill

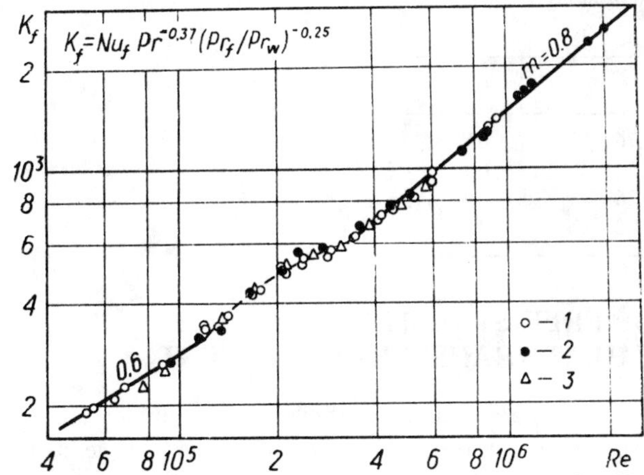

FIG. 7.8 Average heat transfer behavior of a cylinder: 1, $d = 30.7$ mm in water; 2, $d = 50$ mm in water; 3, $d = 32$ mm in air.

and Bernstein [51] suggest the following single relationship, which represented the data over wide ranges of Re:

$$\mathrm{Nu}=0.3+\frac{0.62\,\mathrm{Re}^{0.5}\,\mathrm{Pr}^{0.33}}{(1+0.41\cdot\mathrm{Pr}^{0.66})^{0.25}}\left[1+\left(\frac{\mathrm{Re}}{28200}\right)^{m}\right]^{k_1} \tag{7.23}$$

where $m = 0.625$ and $k_1 = 0.8$ for $\mathrm{Re} < 4000$, and $m = 0.5$ and $k_1 = 1$ for $\mathrm{Re} > 4000$. However, these more complex equations are probably too complex for practical application in industry and design.

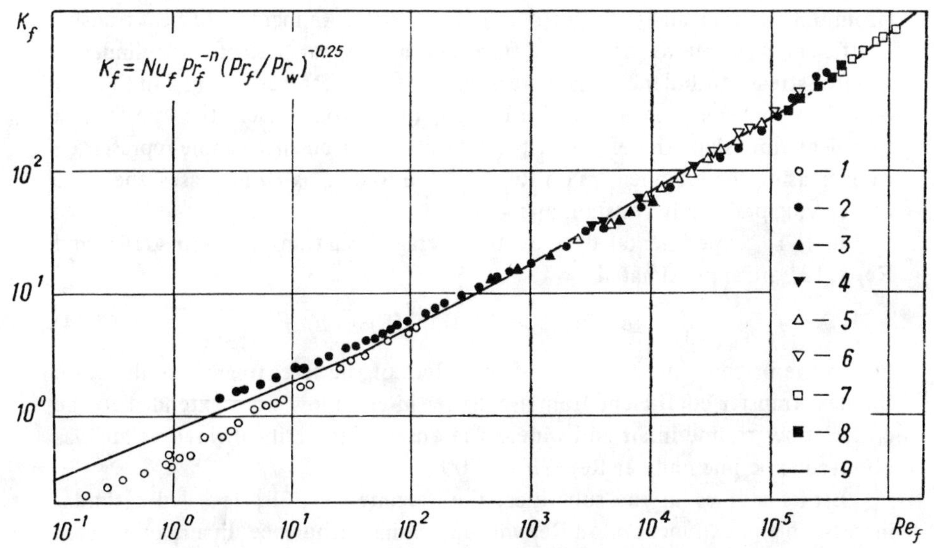

FIG. 7.9 Comparison of data for heat transfer between a cylinder and fluids: 1, [5]; 2, [12]; 3, [29]; 4, [13]; 5, [34]; 6, [16]; 7, [26]; 8, [35]; 9, the present work.

TABLE 7.1 Values of c and m in Eq. (7.5) as a Function of Re Range

Re	c	m
1-40	0.76	0.4
40-1000	0.52	0.5
$1 \cdot 10^3 - 2 \cdot 10^5$	0.26	0.6
$2 \cdot 10^5 - 2 \cdot 10^6$	0.023	0.8

7.4 THE EFFECT OF FREE-STREAM TURBULENCE ON THE AVERAGE HEAT TRANSFER COEFFICIENT

All the test data on the average heat transfer coefficient for a cylinder in crossflow, presented in the previous sections, were determined with at least 1% turbulence in flows of air and liquids.

It has already been shown that the local heat transfer of a circular tube is considerably enhanced by an increase of turbulence in the free stream. In the subcritical flow regime, the free-stream turbulence may be a cause of turbulization in the boundary layer and of an earlier onset of the critical flow regime. The average value of the heat transfer is also affected by the free-stream turbulence.

A more detailed consideration of the effect of the free-stream turbulence on average heat transfer behavior of a cylinder in crossflow is given in references [33, 34].

In the study by Epik and Kozlova [110], average heat transfer coefficients for a cylinder in crossflow was studied in air with a wide range of turbulence (from 0.3 to 23%), and over $1 \cdot 10^3 > \mathrm{Re}_f > 1 \cdot 10^5$. An increase of heat transfer coefficient was found both in the front and in the rear part of the cylinder as the free-stream turbulence was increased, but the effect was minimal in the rear part. It is convenient to use the form of Eq. (7.21) to represent the transfer in a turbulent flow since the effect of the free-stream turbulence can be represented by increasing the exponent m on Re_f from 0.6 to 0.65 as Tu increases, the value of c staying practically constant at $c = 0.26$.

In [34], experimental data on the average heat transfer at $Tu < 14\%$ and $\mathrm{Re}_f < 10^4$ are approximated by

$$\mathrm{Nu}/\mathrm{Nu}_{Tu=0} = 1 + 0.09\,(\mathrm{Re} \cdot Tu)^{0.2} \tag{7.24}$$

In the present study, studies of the effect of the free-stream turbulence on the heat transfer coefficient from a cylinder in crossflow were extended to the critical flow regime in air and water. The onset of the critical flow regime was observed experimentally at $\mathrm{Re}_f \cdot Tu \geqslant 1500$.

Earlier studies in the subcritical flow regime [37, 38] revealed a similar increase in the exponent m on Re_f with increasing turbulence. In air, an increase in turbulence up to 15% is reflected by an increase of the value of m up to 0.63 (Fig. 7.10). In the subcritical range of Re_f, an increase of 35–38% in heat transfer coefficient occurs as Tu increases from 1.2 to 15%. At higher values of Re_f, the effect of Tu on the average heat transfer is even more pronounced. The exception is in a small range of Re_f at the beginning of the critical flow regime. In the supercritical flow regime, at Re_f up to $1 \cdot 10^6$, an intensive augmentation of the heat transfer is observed, and m approaches 0.8. Here an increase in Tu from 1.2 to 15% corresponds to a 55% augmentation of heat transfer, versus 38% for a similar growth of turbulence in the subcritical flow.

Figure 7.11 presents experimental data, as a first approximation, in the form of Eq. (7.25), which contains an auxiliary factor $Tu^{0.15}$. Good agreement with the test points is noted for different Tu for both the critical and the supercritical flows of air.

In this way, a general relation for $Tu > 1\%$ in the subcritical flow of air may be suggested:

$$\mathrm{Nu}_f = 0.23\,\mathrm{Re}_f^{0.6} \cdot Tu^{0.15} \tag{7.25}$$

and the corresponding relation for the supercritical regime is

$$\mathrm{Nu}_f = 0.02\,\mathrm{Re}_f^{0.8} \cdot Tu^{0.15} \tag{7.26}$$

In water, similar studies were performed using a cylinder of 30.1 mm diameter, with Re_f from $5 \cdot 10^4$ to $1.4 \cdot 10^6$, with a variation of the free-stream velocity from 0.9 to 15.5 m/s and with Tu from 1 to 10%.

In the critical flow regime, the average value of the heat transfer coefficient (Fig. 7.12) is more sensitive to changes of Tu in the upper range (4 to 9%). In

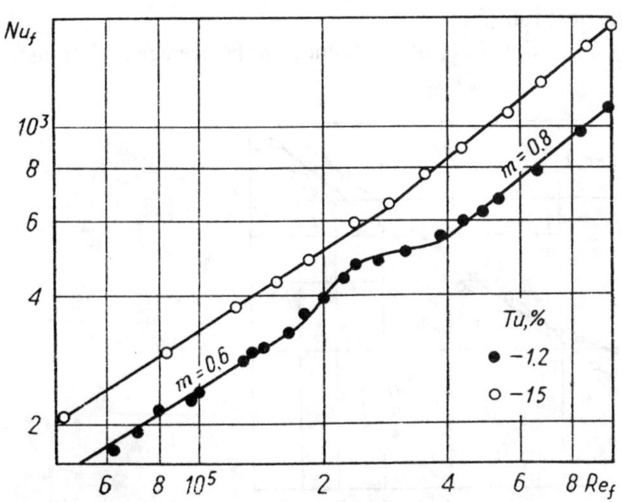

FIG. 7.10 Average heat transfer coefficient as a function of free-stream turbulence in air.

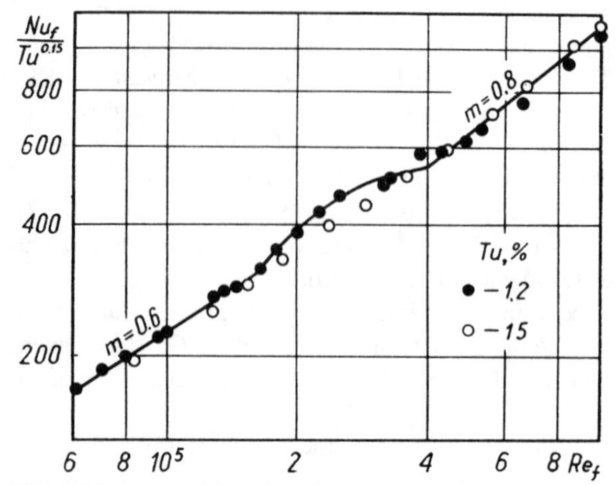

FIG. 7.11 A general function of the heat transfer in relation to free-stream turbulence in air.

the supercritical flow regime, the augmentation of the heat transfer is less pronounced for an increase of *Tu* from 1 to 7%.

As is commonly accepted, in liquid flows (Pr > 1) the thermal resistance is concentrated in the viscous sublayer of the turbulent boundary layer, and ordinary turbulent fluctuations are intensive enough to penetrate to the cylinder wall and to cause a more or less pronounced effect on the heat transfer.

There are several ways of accounting for the effect of the free-stream turbulence on the heat transfer (see Chap. 5). One of these is to use an exponential relation for the average heat transfer.

Judging from Fig. 7.13, the effect of turbulence on the average heat transfer

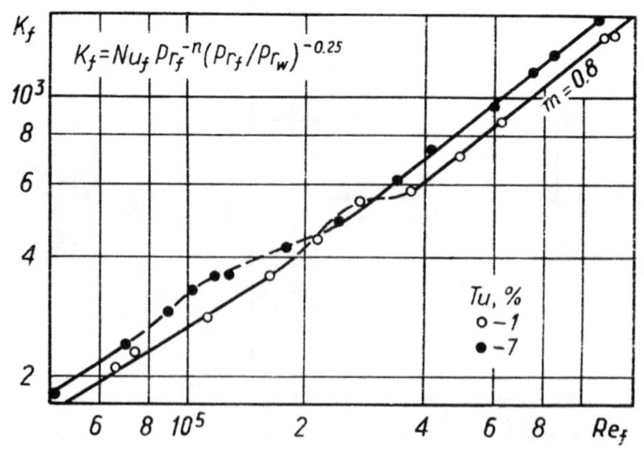

FIG. 7.12 Average heat transfer coefficient as a function of free-stream turbulence in water.

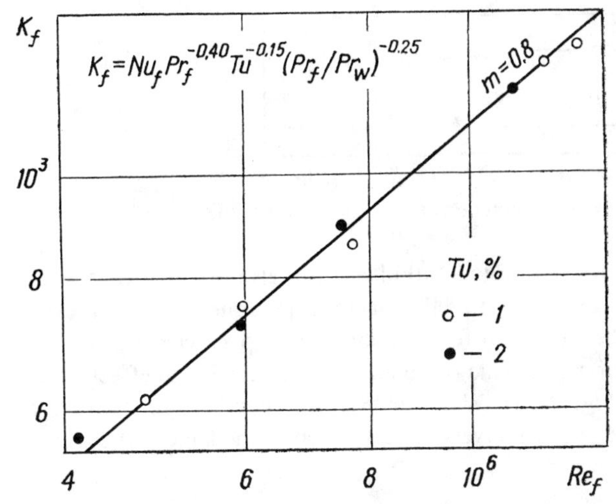

FIG. 7.13 A general function for heat transfer versus free-stream turbulence in water for the supercritical range of Re_f.

can be accounted for by using an auxiliary factor $Tu^{0.15}$. Experimental data for the supercritical flow regime are satisfactorily approximated by

$$Nu_f = 0.023\, Re_f^{0.8} \cdot Pr_f^{0.4} \cdot Tu^{0.15} (Pr_f/Pr_w)^{0.25} \tag{7.27}$$

A comparison (Fig. 7.14) of our results on the effect of turbulence with the results of other authors gave a satisfactory agreement for the various ranges of Re_f.

Equation (7.27) with $Tu^{0.15}$ provides a good approximation for the average heat transfer coefficient including the effect of the free-stream turbulence.

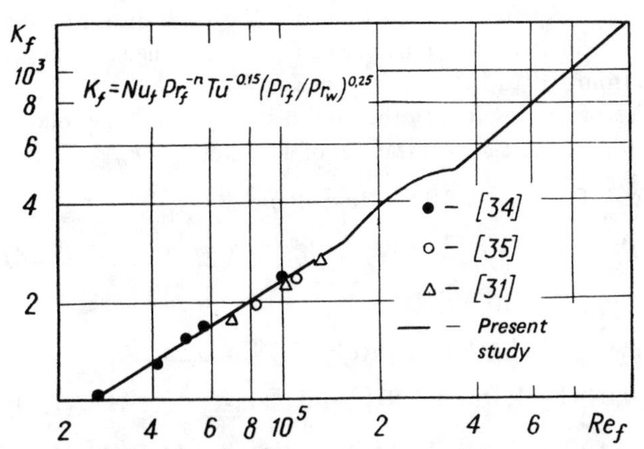

FIG. 7.14 A comparison of different studies of the average heat transfer of cylinders in turbulent flows.

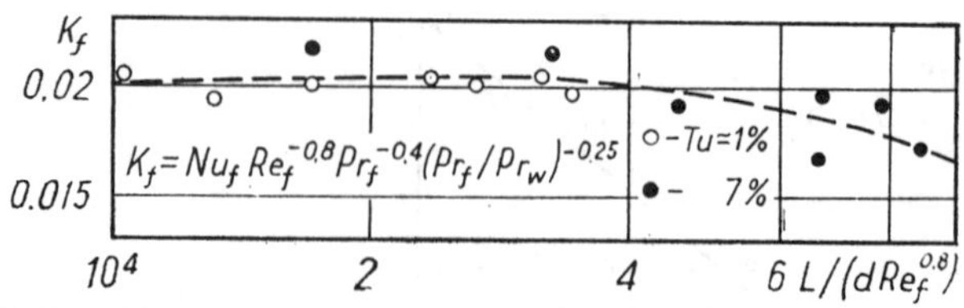

FIG. 7.15 Average heat transfer as a function of macro scale of turbulence.

The large length scale (L) of the turbulence can also be important. From our measurements [38], an increase of the macro scale in the free stream exerts a complex effect on the heat transfer of a cylinder. Up to certain values of Re_f and L, an increase of macro-scale turbulence has practically no effect on the average heat transfer (Fig. 7.15). With further growth of L, an increase of macro-scale turbulence exerts an adverse effect, because the large vortices cause a decrease of the heat transfer.

7.5 THE BLOCKAGE FACTOR AND THE AVERAGE HEAT TRANSFER COEFFICIENT

In a number of practical implementations, circular cylinders and tubes operate in flows limited by channel walls, with considerable blockage factors. A higher blockage factor leads to a higher velocity outside the boundary layer, and to different distributions of pressure and velocity on the cylinder circumference.

Consequently, changes of the shell–cylinder distances lead to changes in the local average heat transfer.

A detailed study of the effect on heat transfer from variation of the blockage factor in the range 0 to 0.827 was performed by Isataev and Akylbaev [39, 129] in air and liquids for Re from 10^4 to $1.6 \cdot 10^5$. They also measured the heat transfer of a cylinder in water and ethylene glycol in a rectangular channel with $k_q = 0.801$ at Re from 10^2 to 10^4 and Pr from 3 to 80.

Taking account also of the results of other authors, they found the following empirical relations for the average heat transfer of a cylinder for $k_q < 0.6$:

$$Nu_f = 0.41\,(1 + 1.18\,k_q^3)\,Re_f^{0.5} \cdot Pr_f^{0.37} + \\ + 0.04\,(1 + 1.18\,k_q^3)^{1.4}\,Re_f^{0.7} \cdot Pr_f^{0.37} \tag{7.28}$$

and for $k_q > 0.6$:

$$Nu_f = 0.41\,(1 + 1.18\,k_q^3)\,Re_f^{0.5} \cdot Pr_f^{0.37} + \\ + 0.04\,(1 + 1.18\,k_q^3)^{1.4}\,Re_f^{0.7} \cdot Pr_f^{0.37} \tag{7.29}$$

The relations apply to Re_f from 10^2 to $2 \cdot 10^5$, Pr_f from 0.7 to 80, and k_q from 0 to 0.9.

In Eqs. (7.28) and (7.29), the effect of the blockage factor is expressed by the value of Re multiplied on a factor $(1 + 1.18k_q)$. Consequently, the account of the blockage factor can also be made by a correction to the relation for the average flow velocity:

$$U^* = U_\infty (1 + 1.18\, k_q^3)^2 \qquad (7.30)$$

In their experiments with water and ethylene glycol, Perkins and Leppert [41] referred the Reynolds number to the average flow velocity,

$$U^* = U_\infty \left(1 - \frac{\pi}{4}\frac{d}{h}\right)^{-1} \qquad (7.31)$$

and suggested the following relationship for the average heat transfer:

$$\mathrm{Nu}_f = (0.3\,\mathrm{Re}_f^{0.5} + 0.1\,\mathrm{Re}_f^{0.67})\,\mathrm{Pr}_f^{0.4}\left(\frac{\mu_f}{\mu_w}\right)^{0.25} \qquad (7.32)$$

for Re_f from $2 \cdot 10^3$ to $1.2 \cdot 10^5$ and $k_q = 0.21$ and 0.415.

In a study of the blockage factor, Gimbutis and Shapola [130] suggested an alternative relationship for the reference velocity,

$$U^* = U_\infty (1 - k_q)^{-0.5} \qquad (7.33)$$

and found accurate results for the average heat transfer and the local heat transfer at the front stagnation point.

Our program included a study of the effect of the blockage factor in the critical and the supercritical flow regime [38]. The experimental values for local heat transfer were integrated along the circumference for k_q from 0.28 to 0.70, and the average heat transfer coefficient was determined for Re_f from $4 \cdot 10^4$ to $1 \cdot 10^6$, taking account of the free-stream velocity. The data fell on a given curve for each respective value of the blockage factor (Fig. 7.16).

With high values of k_q, the velocity outside the boundary layer increases up to the mid cross section, so that neither the free-stream velocity U_∞ nor the velocity in the least free space U can reflect the actual fluid dynamics. Thus, an appropriate value of the reference velocity must be sought between U_∞ and U.

The interpretation in Fig. 7.17 involves a correction $(1 - k_q^2)$ to the velocity in the least free space, so that the reference velocity is

$$U^* = (1 - k_q^2)\, U \qquad (7.34)$$

Our results for the supercritical flow regime with different blockage factor were approximated by

$$\mathrm{Nu}_f = 0.023\,[(1 - k_q^2)\,\mathrm{Re}_f]^{0.8}\,\mathrm{Pr}_f^{0.4}\,(\mathrm{Pr}_f/\mathrm{Pr}_w)^{0.25} \qquad (7.35)$$

With high levels of turbulence, Eq. (7.35) should also include a term Tu to a given power.

For infinite flows, when $k_q = 0$, Eq. (7.35) coincides with Eq. (7.22).

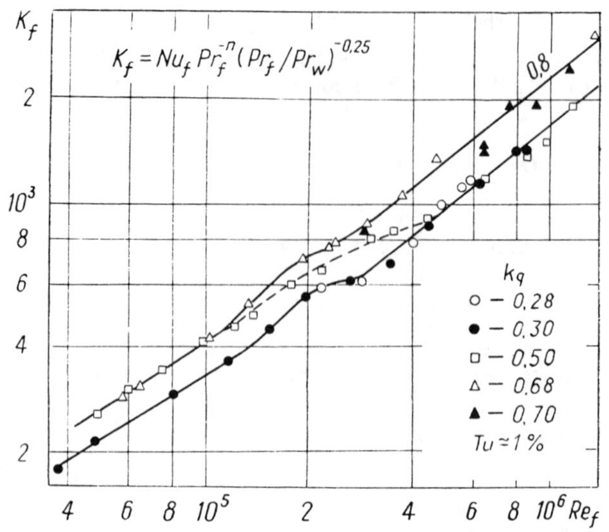

FIG. 7.16 Average heat transfer coefficient as a function of blockage factor.

7.6 THE AVERAGE HEAT TRANSFER OF A ROUGH-SURFACE CYLINDER

Our heat transfer measurements on rough-surface cylinders [131, 132] in water flow in the range of Re_f from $4 \cdot 10^4$ to $1.6 \cdot 10^6$ (Fig. 7.18) revealed that there was an augmentation of average heat transfer coefficient with increase of the

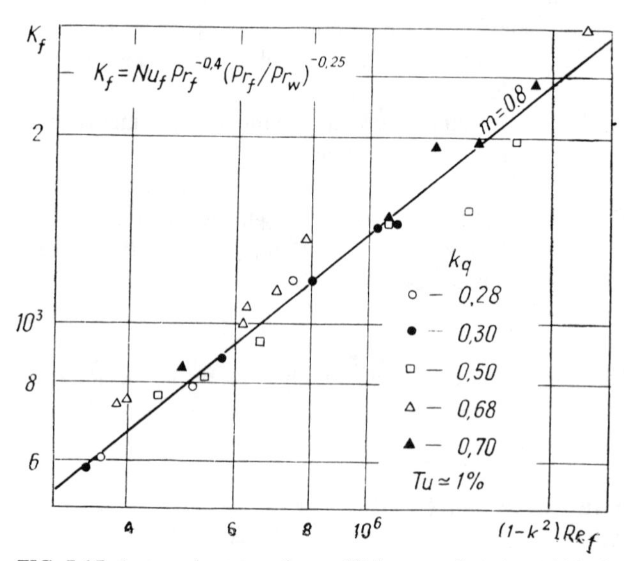

FIG. 7.17 Average heat transfer coefficient as a function of blockage factor.

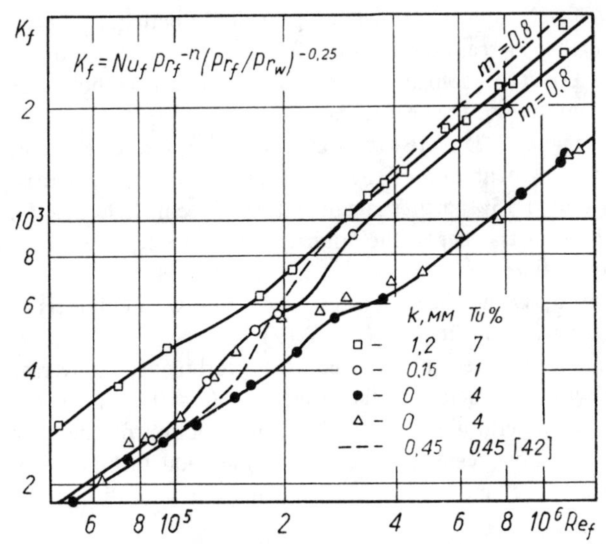

FIG. 7.18 Average heat transfer of a rough-surface cylinder as a function of the turbulence level.

relative size of the surface elements, and an earlier onset of the critical flow regime. Similar results for air were found by Achenbach [42].

In the subcritical flow regime, and with moderate height of the surface elements ($k = 0.15$ mm), the heat transfer augmentation relative to a smooth cylinder is not significant (Fig. 7.18). The augmentation is caused by the enhancement of velocity fluctuations in the laminar boundary layer caused by the surface elements. As Re_f is increased, a critical value of the surface roughness parameter $k^+ = u_* k/\nu$ is reached and laminar–turbulent transition occurs in the boundary layer. For a rough-surface cylinder, Re_f is also a function of the surface roughness, and depends on the average friction drag of the cylinder. An increase of Re_f leads to an increase of the value of the friction velocity u_* and of the surface roughness parameter k^+, leading to transition. This is the opposite of what is observed on a rough-surface plate.

On a rough-surface cylinder ($k = 0.15$ mm) the critical flow regime is established at $Re_f = 10^5$, compared with $Re_f = 2 \cdot 10^5$ for a smooth cylinder. The early critical regime also includes the separation bubble, as manifested by the deviation from the subcritical curve in Fig. 7.18. After the disappearance of the separation bubble on a rough cylinder at $Re_f = 2.3 \cdot 10^5$ and on a smooth cylinder at $Re_f = 4 \cdot 10^5$, the curve of heat transfer enters the supercritical flow regime.

The heat transfer behavior of a rough-surface cylinder in the subcritical flow regime, and in the range of existence of the separation bubble, is similar to that of a smooth cylinder at $Tu = 4\%$. This shows the additive action of the free-stream turbulence of 1% and the higher roughness-generated fluctuations in this range of Re_f. With a further increase of Re_f and after the disappearance of the separation bubble, the effect of a higher roughness and a thinner boundary layer gives greater augmentation of the heat transfer.

In the critical flow regime (Fig. 7.18) for water ($\mathrm{Pr}_f \approx 6$), surface elements of 0.15 mm give an 85% augmentation of the heat transfer. A further increase of k to 1.2 mm gives only a 15% further augmentation. We conclude that for the augmentation of the heat transfer in the critical flow of water, a moderate surface roughness is most efficient.

Analogous relations for average heat transfer were found with cylinders of different rough surfaces in water at $Tu = 7\%$ (Fig. 7.19).

Figure 7.20 illustrates the effect of the free-stream turbulence on average heat transfer behavior with a moderate surface roughness ($k = 0.15$ mm). The curve may be of help in determining the predominant fraction of the total turbulence. In the critical flow regime, the effect of the total turbulence of $Tu = 1\%$ on a rough-surface cylinder is similar to that on a smooth cylinder for $Tu = 4\%$.

The level of turbulence in the critical range of Re_f is similar to that corresponding to a surface roughness of $k = 0.15$ mm, but the trend of the curve is typical of the effect of the free-stream turbulence. We conclude that the fluid dynamics and heat transfer are, in this situation, governed by the free-stream turbulence.

In our studies on smooth cylinders in the critical flow regime, no augmentation of the heat transfer was observed with the increase of the free-stream

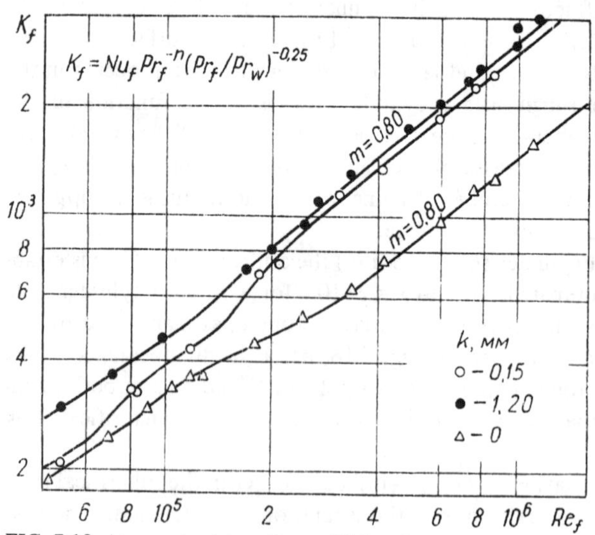

FIG. 7.19 Average heat transfer coefficient for a rough-surface cylinder at $Tu = 7\%$.

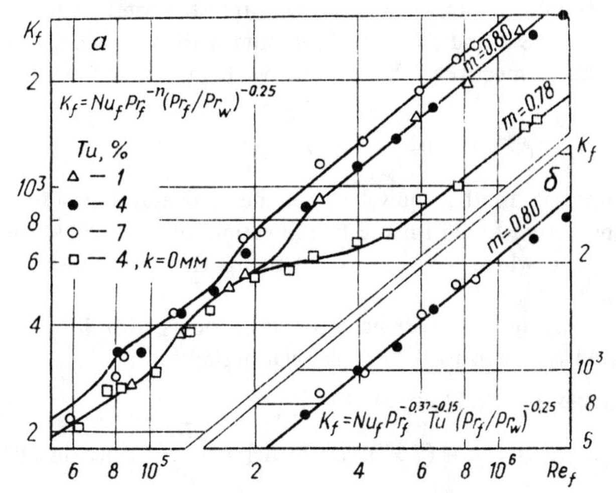

FIG. 7.20 Average heat transfer behavior of a rough-surface cylinder for $k \approx 0.15$ mm: α, without and δ, with account of the turbulence level.

turbulence from 1 to 4%. By analogy, the free-stream-induced component of the total turbulence must be constant on a rough-surface cylinder, for the range of free-stream turbulence from 1 to 4%. For a constant size of the surface elements ($k = 0.15$ mm), the roughness-induced component of turbulence is also constant. Thus a change of the free-stream turbulence from 1 to 4% has only a small effect on the heat transfer of a rough-surface cylinder, just as it has on a smooth cylinder (Fig. 7.20).

An analysis of the experimental results for smooth cylinders in water revealed a 20% augmentation of the heat transfer with the increase of turbulence from 4 to 7%. An augmentation was observed for the rough-surface cylinder as Tu was increased from 4 to 7%; this is similar to the effect observed for a smooth cylinder for $Tu \geqslant 1\%$, the exponent on Tu being 0.15.

7.7 AVERAGE HEAT TRANSFER BEHAVIOR OF CYLINDERS WITH NONCIRCULAR CROSS SECTION

Tubes in industrial heat exchangers may be of noncircular cross section, and this introduces considerable complication into the calculation of their average heat transfer. The difficulty lies in assessing the effect of their profile, or geometry. Numerous authors have performed measurements of the heat transfer coefficient from different cylinders in air [50, 133]. To our knowledge, the present results are the first that have been obtained for elliptic cylinders for other fluids.

Figure 7.21 presents our results for average heat transfer coefficient for elliptic cylinders with $d_1 = 58.3$ and $d_2 = 30.3$ mm and with flows along the major or the minor axes, respectively. The results were interpreted by the equation

$$\mathrm{Nu}_{fd_e}/\mathrm{Pr}_f^{0.37}(\mathrm{Pr}_f/\mathrm{Pr}_w)^{0.25} = f(\mathrm{Re}_{fd_e}) \tag{7.36}$$

and the values were referred to the equivalent channel diameter $d_e = 4F/p$. Different Nusselt numbers were found for the two directions of the flow. Much better agreement was found when the axis parallel to the flow was chosen as the characteristic dimension.

In Fig. 7.22 the average heat transfer behavior of cylinders with different profiles was approximated satisfactorily by a single relationship,

$$\mathrm{Nu}_{fd_1} = 0.27\,\mathrm{Re}_{fd_1}^{0.6} \cdot \mathrm{Pr}_f^{0.37}(\mathrm{Pr}_f/\mathrm{Pr}_w)^{0.25} \tag{7.37}$$

The results for constant q_w, Eq. (7.37), are higher than for constant t_w, Eq. (7.21).

The use of the axis parallel to the flow as the characteristic dimension is only possible for cylindrical geometries.

At present, the various authors are not unanimous about the choice of characteristic dimension. The various publications suggest the following values: equivalent diameter, width normal or parallel to the flow, square root of the surface area, etc. None of them can be suggested as a unique reference size, and there exists no general equation to serve for the heat transfer approximation for different geometries.

On tubes of different profiles, the regions of the laminar, turbulent, and vortex flows occupy different fractions of the circumference. This must be

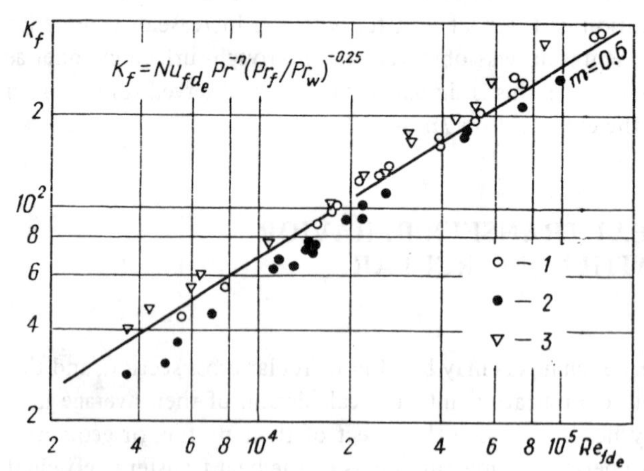

FIG. 7.21 Heat transfer behavior of curvilinear bodies in different fluids: 1, circular cylinder; 2, 3, elliptic cylinders with flows parallel to the major and minor axes, respectively.

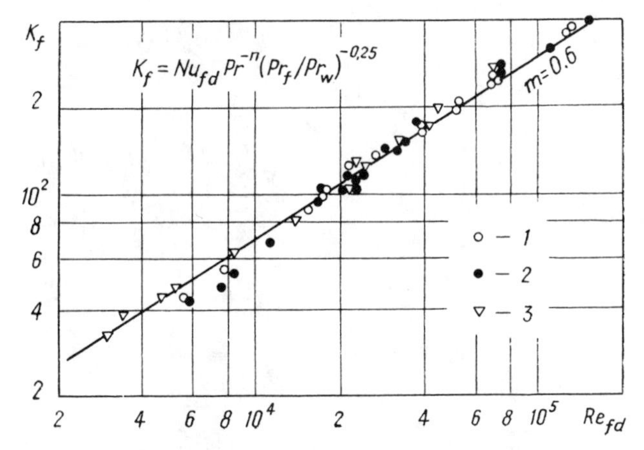

FIG. 7.22 Heat transfer from curvilinear bodies in different fluids: 1, 2, 3, as in Fig. 7.21.

reflected by different exponents on Re and Pr in the corresponding relations. We conclude, therefore, that a unique geometrical parameter is unlikely to exist although one or another of the values can be used in approximate engineering calculation.

An alternative approach to the heat transfer calculation for complex shapes may be possible, based on the hydraulic drag, which depends on the geometry in separated flows. Such an approach is one of the future prospective ways of solving the heat transfer problem of different bodies in crossflow.

Heat Transfer Behavior of Two Cylinders in Series

Table 14 in the Appendix shows significant changes of the average heat transfer of both the leading and the rear cylinder, when $x/d \leqslant 3$, where x is the distance between the cylinder centers. A small effect persists up to $x/d = 6$. Thus, the average heat transfer coefficient tends toward a constant value with the onset of the regular vortex shedding.

The heat transfer of the leading cylinder is similar to that of a single one. For the average heat transfer coefficient on the rear cylinder, the following expression is suggested for the subcritical flow regime,

$$Nu_f = 0.35\, Re_f^{0.6}\, Pr_f^{0.37}\, (Pr_f / Pr_w)^{0.25} \tag{7.38}$$

and

$$Nu_f = 0.17\, Re_f^{0.75}\, Pr_f^{0.37}\, (Pr_f / Pr_w)^{0.25} \tag{7.39}$$

for the critical flow regime at $Re_f = 3 \cdot 10^5$ to $1.4 \cdot 10^6$.

The values here were referred to the velocity in the least free space. Heat transfer between the rear cylinder and the fluid is augmented by the nonisotropic highly turbulent flow. The relations are in good agreement with those given in reference [47].

CHAPTER

EIGHT

CONCLUSION AND GENERAL REMARKS

The preceding chapters presented experimental and analytical results on the fluid dynamic and heat transfer behavior for cylinders in crossflow, including some data on elliptic cylinders. The measured values gave the distributions of pressure, shear stress, and hydraulic drag on the cylinder surface and also local and average heat transfer coefficients. The measurements were performed over a wide range of free-stream turbulence and channel blockage factor.

Chapter 8 contains general remarks and practical suggestions on the determination of hydraulic drag and heat transfer on cylinders in different fluid flows with complex boundary conditions.

8.1 THE BOUNDARY LAYER AND LOCAL HEAT TRANSFER

A close relationship is observed between the development of the boundary layer on the one hand and the heat transfer from the cylinder surface on the other. Both experiments and analytical solutions indicate that there is a laminar boundary layer on the front part of a cylinder. As the thickness of this boundary layer increases with angular distance, thermal resistance increases, and the heat transfer decreases (see Chaps. 4 and 6). An exception is observed for flows with high blockage factor ($k_q > 0.5$); here, due to the high pressure gradient, the boundary layer becomes thinner, and the heat transfer higher, as the separation point is approached.

The measurements indicate that there is not a steady-state boundary layer downstream of the separation point. The heat transfer in the rear part is governed by a vortex-separated flow.

The vortex processes depend on the conditions in the near wake, and on the regularity or irregularity of the separation itself, which in turn depends on the flow regime operating. The considerably lower heat transfer coefficients observed in the rear part in the critical flow regime are connected with the absence of a regular vortex shedding. This is reflected in changes in the average heat transfer coefficient.

The critical flow regime is distinguished by a number of specific features. The curves of the local heat transfer have a second minimum, corresponding to the separation of the turbulent boundary layer. The first minimum corresponds to the laminar-turbulent transition in the boundary layer. Depending on the specific values of Re and *Tu*, the transition occurs either on the surface or in the separation bubble.

We maintain, therefore, that any consideration of the heat transfer behavior of a cylinder in crossflow must be closely related to the fluid dynamic behavior.

8.2 FLUID DYNAMICS ON A CYLINDER

The fluid dynamics of flow over a cylinder are governed by the distribution of pressure and velocity over the surface. Outside the boundary layer, a relation between the two distributions is clearly observed:

$$U_\varphi = U_\infty \sqrt{1 + \frac{2(p_\infty - p_\varphi)}{\rho U_\infty^2}} \tag{2.12}$$

The various flow regimes are reflected in the curves (Fig. 8.1) for pressure distribution over a cylinder. Curve 1 represents the analytical expression for the

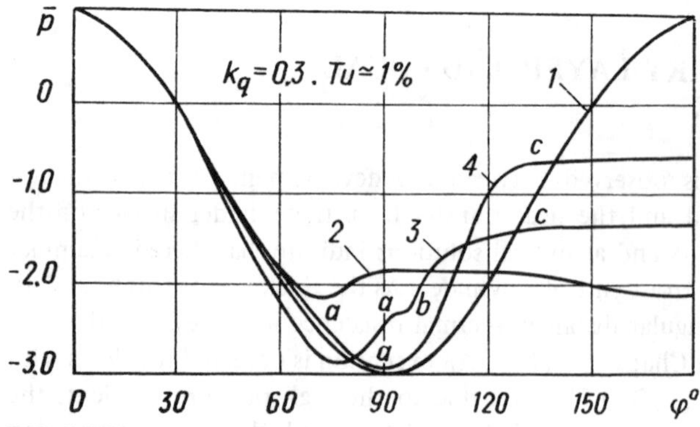

FIG. 8.1 Pressure distribution on a circular cylinder.

pressure distribution over a cylinder in crossflow of an ideal fluid, when both friction and viscosity are ignored. It shows a symmetric distribution of pressure with maxima in the front and rear stagnation points, and a minimum in the middle cross section.

Curve 2 represents the distribution of pressure on a cylinder surface in water at moderate values of Re in the subcritical flow regime. Under the action of the viscous forces, a laminar boundary layer is formed on the cylinder, its thickness increasing downstream. Up to $\varphi = 40°$ (Fig. 8.1), no effect of the Reynolds number is noted on the distribution of pressure and, consequently, on the velocity outside the boundary layer. The pressure distribution is in good agreement with the curve for ideal fluid. At $\varphi > 40°$, and with real fluids the effect of Re becomes significant. The kinks in curve 2 at $\varphi \approx 80°$ and at *a* correspond to the separation of the boundary layer and to the beginning of a complex vortex flow in the rear part of the cylinder.

With an increase of the Reynolds number ($Re > 1.5 \cdot 10^5$), the critical flow regime is established. On the front part of the cylinder, separation of the laminar boundary layer and the separation bubble are observed. These are followed by the reattachment of the boundary layer, the laminar–turbulent transition, and a turbulent boundary layer on the rear part of the cylinder. The laminar–turbulent transition in the boundary layer absorbs additional amounts of kinetic energy from the turbulent fluctuations, and the velocity gradient on the surface increases considerably. As a result, the turbulent boundary layer can resist the effect of the increased velocity gradient in the free stream ($dp/dx > 0$). The boundary layer is extended downstream and separates at $\varphi \approx 140°$.

Curve 3 in Fig. 8.1 represents the distribution of pressure on the surface of a cylinder in a crossflow of water at $Re \approx 4.5 \cdot 10^5$. Here *a* corresponds to the separation of the laminar boundary layer and to the formation of the separation bubble, *b* to the reattachment, and *c* to the separation of the turbulent boundary layer. A further increase of Re leads to the supercritical flow regime, characterized by the absence of the separation bubble. The laminar–turbulent transition occurs inside the boundary layer, which is still extended downstream and separates at $\varphi \approx 140°$.

In the curve for the pressure distribution in air at $Re_f = 10^6$ (Fig. 8.1, curve 4), *a* is the laminar–turbulent transition in the boundary layer, and *c* is the separation point of the turbulent boundary layer. With a further increase of Re, combined with an increase of turbulence (*Tu*), the laminar–turbulent transition in the boundary layer is moved upstream, and the point of the turbulent boundary layer separation is almost stationary.

Figure 8.2 presents relations for the location of the points of the laminar boundary layer separation and of the laminar–turbulent transition in the boundary layer for different values of Re_f. The curves were compiled from measurement of the fluid dynamics parameters ($\bar{\tau}$, $\bar{p}$) and of the heat transfer. They

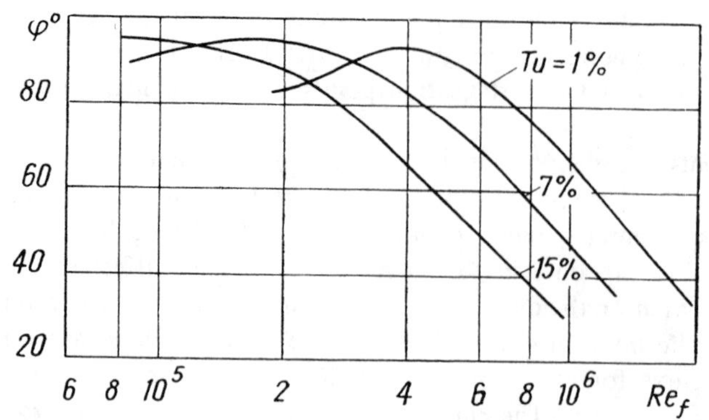

FIG. 8.2 Curves for the location of the separation point and of the transition point for different Re_f and *Tu*.

cover the range of Re_f from 10^4 to $2 \cdot 10^6$. With an increase of Re_f, the transition point is removed downstream and reaches $\varphi \approx 95°$, giving rise to the separation bubble. With the disappearance of the separation bubble, the separation point is moved back upstream by an amount that depends on Re_f and *Tu*. The effect of the blockage factor is analogous; the separation point is moved to 130°, versus 140° for a lower blockage factor.

The processes of both heat transfer and fluid dynamics are governed by the distribution of velocity outside the boundary layer. Flow velocity in the front part of the cylinder can be determined by the relation of Hiemenz [Eq. (3.6)], which assumes infinite flow ($k_q = 0$). However, in industrial applications flows over bodies include the effect of the channel walls and thus of the blockage factor. The latter factor exerts its effect via the velocity distribution outside the boundary layer. Our analysis of numerical experimental data resulted in a modified relationship,

$$\frac{U_\varphi}{U_\infty} = A_1^* \frac{x}{d} + A_2^* \left(\frac{x}{d}\right)^3 + A_3^* \left(\frac{x}{d}\right)^5 \tag{8.1}$$

for the velocity distribution on the front part of the cylinder, taking account of the blockage factor. The values of the constant $A^* = f(k_q)$ are given in Fig. 8.3.

8.3 HYDRAULIC DRAG

The fluid dynamic phenomena in flow over a cylinder are closely related to hydraulic drag on the cylinder. The total hydraulic drag of a cylinder consists of the sum of the friction drag P_f and the pressure drag P_d, acting on the surface. A dimensionless expression for the total hydraulic drag is

$$c_D = \frac{P_f + P_d}{\frac{\rho U_\infty^2}{2} d \cdot l} \tag{8.2}$$

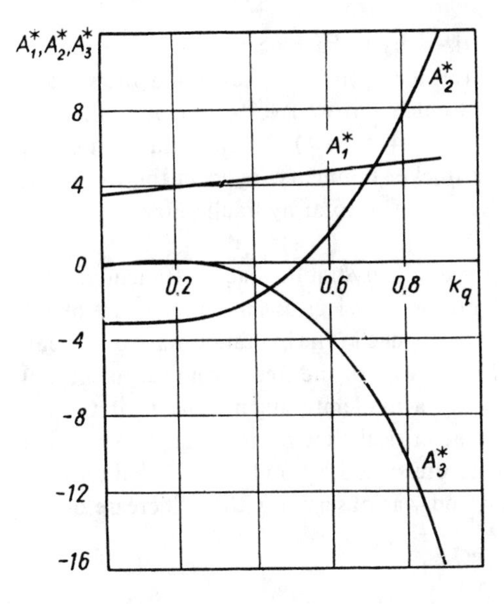

FIG. 8.3 Chart for estimating the constants in Eq. (8.1).

In the low range of Re the friction drag component predominates, but with the increase of Re it decreases, and at Re $> 10^4$ constitutes only 3 to 1% of the total hydraulic drag.

For the lower range of Re (Fig. 8.4, zone 1), where the friction drag predominates, the dimensionless hydraulic drag coefficient c_D is highly dependent on the Reynolds number. Somewhat different tendencies are observed in the whole range of the subcritical flow regime (Fig. 8.4, zone 2). The critical flow

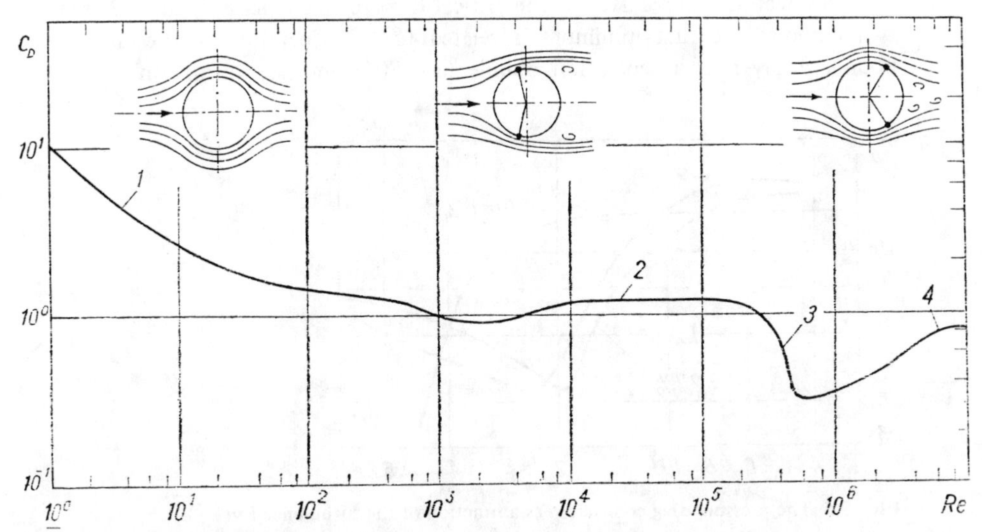

FIG. 8.4 The hydraulic drag coefficient as a function of the Reynolds number.

regime (Fig. 8.4, zone 3) is characterized by a sharp decrease of the total hydraulic drag coefficient, because the turbulent boundary layer and the shift of separation point downstream lead to a much narrower vortex region in the rear.

In the supercritical flow regime (Fig. 8.4, zone 4), the upstream shift of the laminar-turbulent transition implies a thicker boundary layer on the front part of the cylinder, and a wider vortex region. The total hydraulic drag coefficient increases.

Figure 8.5 shows the relation between the hydraulic drag coefficient and the free-stream turbulence. A higher turbulence level gives an earlier onset of the critical flow regime, and a corresponding change in the hydraulic drag coefficient.

In the subcritical flow regime, the increase of the free-stream turbulence, or of the surface roughness, can give only a moderate augmentation of the hydraulic drag coefficient. However, in the critical flow regime (Fig. 8.6), a higher surface roughness k/d leads to a higher pressure drag, and to an earlier onset of the critical flow regime. The same trend was observed with an increase of the turbulence level.

8.4 LOCAL HEAT TRANSFER

Changes of the fluid dynamics are also reflected in the local heat transfer of a cylinder in crossflow. At low Reynolds numbers (Fig. 8.7, curve 1), the heat transfer is at its maximum in the front part, and gradually decreases around the circumference with the increase of the boundary layer thickness. At higher Re (Fig. 8.7, curve 2), a gradual increase of heat transfer is observed in the vortex region, following the boundary-layer separation at $\varphi \approx 80°$.

The heat transfer curve for the critical flow regime (Fig. 8.7, curve 3) has two minima: the first minimum is related to the separation of the laminar boundary layer as a separation bubble ($\varphi = 80°$) and the formation of the

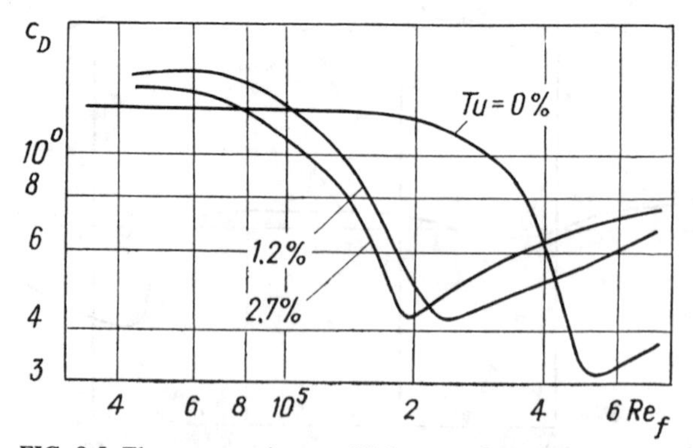

FIG. 8.5 The pressure drag coefficient as a function of the turbulence level.

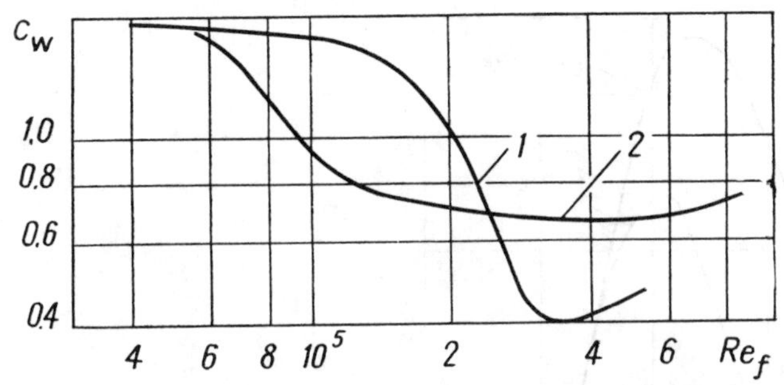

FIG. 8.6 The pressure drag coefficient as a function of the relative roughness at $Tu = 1\%$: 1, $k/d = 0$; 2, $k/d = 5 \cdot 10^{-3}$.

turbulent boundary layer, and the second to the separation of the turbulent boundary layer itself.

In the supercritical flow regime ($\mathrm{Re}_f \approx 2 \cdot 10^6$) (Fig. 8.7, curve 4), the first heat transfer minimum is related to the laminar-turbulent transition in the boundary layer ($\varphi = 40°$), and the second to the separation of the turbulent boundary layer. The exact location of the laminar-turbulent transition is a function of the Reynolds number and of the turbulence level.

For the supercritical flow regime (Fig. 8.8), the first heat transfer minimum is shifted upstream with increased free-stream turbulence, so that the laminar boundary layer region reduces in size.

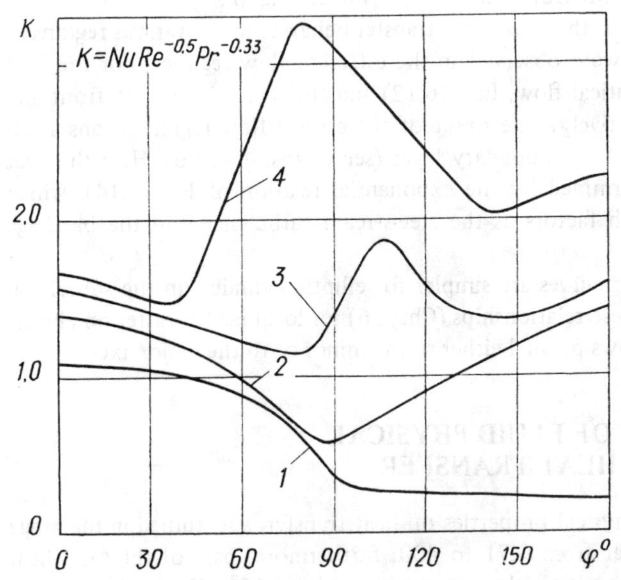

FIG. 8.7 Local heat transfer behavior as a function of angular distance.

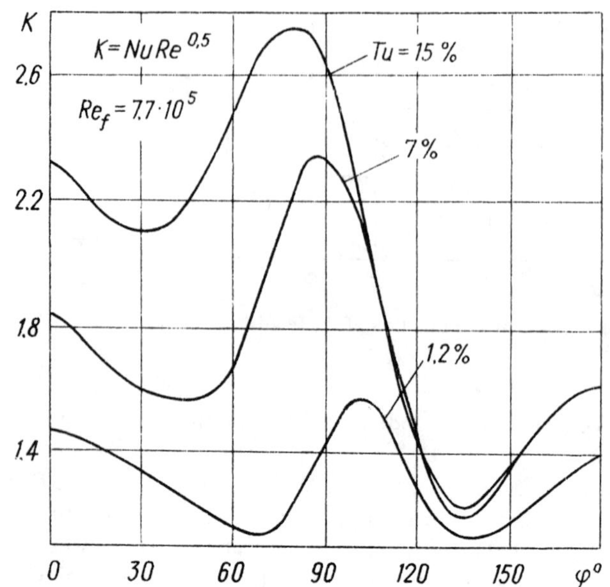

FIG. 8.8 Average heat transfer as a function of turbulence in the supercritical flow regime.

A significant effect of the surface roughness on local heat transfer is also noted. It is more pronounced in the critical flow regime (see Section 6.4).

A higher blockage factor in the channel gives a pronounced augmentation of the heat transfer in the front part of the cylinder (Fig. 6.26).

Specific changes in the local heat transfer behavior in the various regions of the cylinder surface were observed in the different flow regimes. For the heat transfer in the subcritical flow, Eqs. (6.12) and (6.13) apply for the front and the rear parts, respectively. The onset of the critical flow regime means a different development of the boundary layer (see Chaps. 4 and 6). Here the heat transfer can be determined by the exponential relation of Eq. (6.14), which takes account of such factors as the free-stream turbulence and the blockage factor.

The heat transfer curves are simpler for elliptic cylinders in the subcritical flow regime. We suggest relationships (Chap. 6) for local heat transfer on elliptic cylinders, with the flows parallel either to the minor or to the major axes.

8.5 THE EFFECT OF FLUID PHYSICAL PROPERTIES ON HEAT TRANSFER

The effect of fluid physical properties on heat transfer was studied in the range of the Prandtl number from 0.71 to 400; furthermore, part of the local heat transfer studies covered Prandtl numbers up to about 10^3. The analysis of the results yielded a relationship for the heat transfer [Eq. (7.5)] that included a term for Pr raised to a power n. For the subcritical flow regime, $n = 0.37$, and for the critical flow regime, $n = 0.40$.

For flows of moderate temperatures, the values of physical properties used in calculating Nu_f, Re_f, and Pr_f should be estimated at the bulk flow temperature t_f in both gases and liquids. With this approach, the effect of the fluid physical property variations is satisfactorily described in terms of the ratio $(\mathrm{Pr}_w/\mathrm{Pr}_f)^p$ for both wall-to-fluid and fluid-to-wall heat transfer.

For wall-to-fluid heat transfer, the mean value of p is 0.25; for fluid-to-wall heat transfer, $p = 0.20$. For moderate temperature differences, use of $p = 0.25$ satisfies both cases. Figure 8.9 presents the functional relationship

$$K_1 = Nu_{f_1}/\mathrm{Nu}_{f_0} = f(\mathrm{Pr}_f/\mathrm{Pr}_w)^p \tag{8.3}$$

for the effect of the fluid physical properties on the heat transfer of a cylinder. For gases with Pr ≈ constant, $\mathrm{Pr}_f/\mathrm{Pr}_w \approx 1$.

In our studies all the liquids (water, transformer oil, and aviation oil) were nontransparent to infrared radiation, and any effect of radiation was excluded. This does not apply to all liquids used in industrial heat exchangers. Studies with organic liquids–saturated hydrocarbons and derivatives of hexane–performed at the Institute of Physical Chemistry in Kazan [134, 135] showed that nearly 20% of the heat transfer could occur by radiation at 100°C. The effect cannot be ignored in the design of heat exchangers to be used with such liquids.

8.6 DETERMINATION OF AVERAGE HEAT TRANSFER COEFFICIENT

For a cylinder in crossflow, the average heat transfer behavior is governed by the fluid physical properties and by the fluid dynamics. Since the fluid dynamics are governed by the Reynolds number, the present study covered Re from 1 to

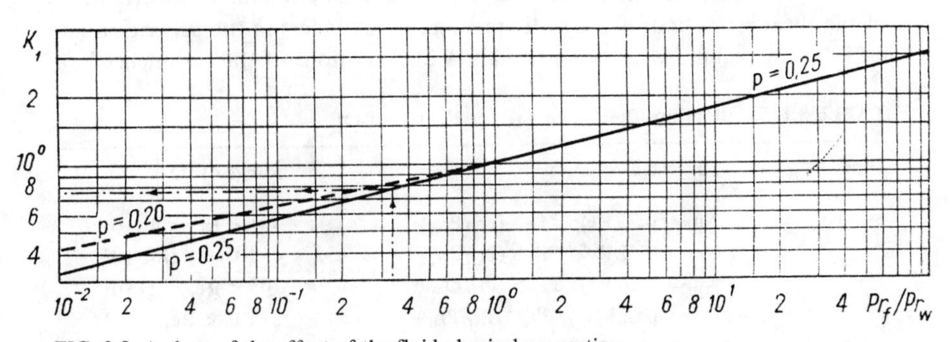

FIG. 8.9 A chart of the effect of the fluid physical properties.

$2 \cdot 10^6$. Both the fluid dynamics and the heat transfer are greatly effected by the turbulence level and by the blockage factor. The study covered Tu up to 15% and k_q up to 0.7.

For $Tu \approx 1\%$, the general relation for the average heat transfer is

$$\mathrm{Nu}_f = c\, \mathrm{Re}_f^m\, \mathrm{Pr}^n\, Tu^{0.15}\, (\mathrm{Pr}_f/\mathrm{Pr}_w)^p \tag{8.4}$$

The endeavor to extract practical conclusions from the experimental results has led to the idea of separate flow regimes in the different ranges of Re_f (Table 8.1 and Fig. 8.10). Thus n, the exponent on Pr_f, is 0.37 for the subcritical flow regime and 0.4 for the supercritical flow regime. The value of m, the exponent on Re_f, varies from 0.4 to 0.8.

In the low range of Re_f, m varies from 0.4 to 0.5 (Fig. 8.10, zone 1), and increases to $m = 0.6$ for the subcritical flow regime (zone 2). In the critical flow regime the curve is uncertain (zone 3), but in the supercritical flow regime, $m = 0.8$.

At low values of Re, account should be taken of free convection using Eqs. (7.8)-(7.13).

When effects of the turbulence level, and/or blockage factor, and/or surface roughness are required, equations from Sections 7.4, 7.5, and 7.6 should be applied.

Figure 8.11 gives Nu_f as a function of Re_f at various Pr_f. When there are considerable temperature differences and/or high turbulence levels, the charts in Figs. 8.9 and 8.12 should be consulted and

$$\mathrm{Nu}_f = \mathrm{Nu}_{f_0} \cdot K_1 \cdot K_2 \tag{8.5}$$

should be applied. Here $K_1 = f(\mathrm{Pr}_f/\mathrm{Pr}_w)$ is found in Fig. 8.9, and $K_2 = f(Tu)$ in Fig. 8.12.

8.7 GENERAL REMARKS

Both experimental and analytical data were used in determining the distributions of the shear stress and of the static pressure on cylinders in the subcritical and critical flow regimes with various fluids. Wide variations of the turbulence level

TABLE 8.1 Relationships Describing Average Heat Transfer

Zones		Suggested equation	Range of Re_f
1	Nu	$\mathrm{Nu}_f = 0.76\, \mathrm{Re}_f^{0.4}\, \mathrm{Pr}_f^{0.37}\, (\mathrm{Pr}_f/\mathrm{Pr}_w)^p$	$10^0 < \mathrm{Re}_f < 4 \cdot 10^1$
2		$\mathrm{Nu}_f = 0.52\, \mathrm{Re}_f^{0.5}\, \mathrm{Pr}_f^{0.37}\, (\mathrm{Pr}_f/\mathrm{Pr}_w)^p$	$4 \cdot 10^1 < \mathrm{Re}_f < 10^3$
3		$\mathrm{Nu}_f = 0.26\, \mathrm{Re}_f^{0.6}\, \mathrm{Pr}_f^{0.37}\, (\mathrm{Pr}_f/\mathrm{Pr}_w)^p$	$10^3 < \mathrm{Re}_f < 2 \cdot 10^5$
4		$\mathrm{Nu}_f = 0.023\, \mathrm{Re}_f^{0.8}\, \mathrm{Pr}_f^{0.4}\, (\mathrm{Pr}_f/\mathrm{Pr}_w)^p$	$2 \cdot 10^5 < \mathrm{Re}_f < 10^7$

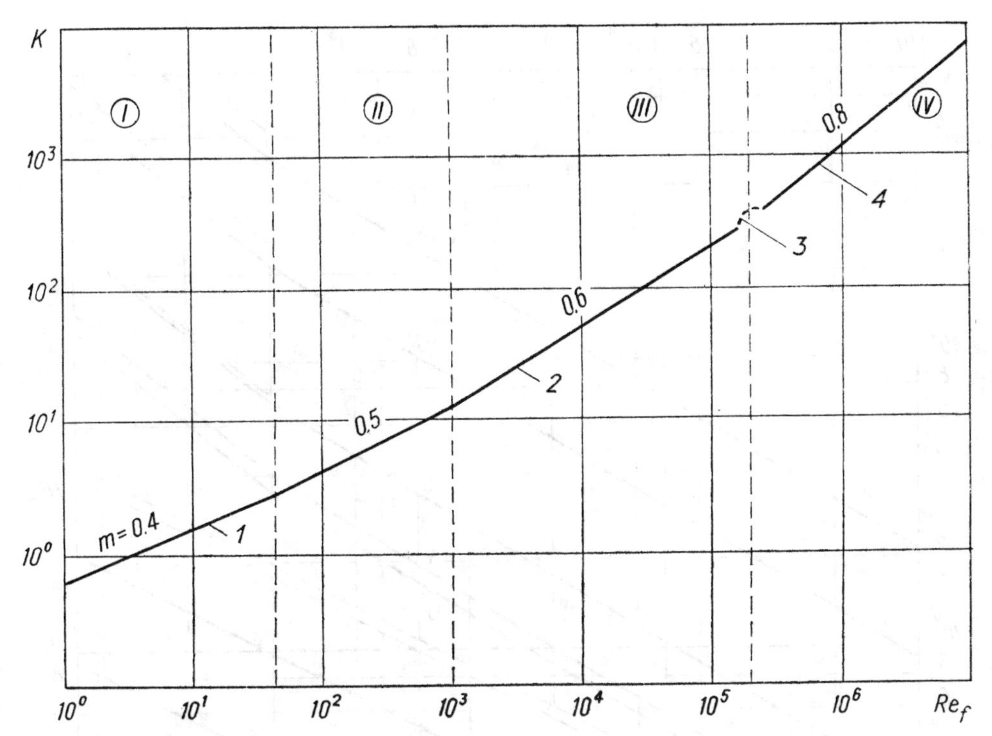

FIG. 8.10 A chart of the average heat transfer.

and of the blockage factor were covered, and the local heat transfer was evaluated. A combined analysis of the fluid dynamical and heat transfer parameters revealed the functional behavior of the laminar-turbulent transition point and the separation point as well as the hydraulic drag and the heat transfer as a function of Reynolds number, turbulence level, and blockage factor. An increase of either the free-stream turbulence or the blockage factor was found to lead to an augmentation of the heat transfer and to a higher friction drag and hydraulic drag in the separate regions on the cylinder surface. Similar effects were observed with the introduction of surface roughness. Analytical solutions show deformations of the velocity and temperature profiles in the boundary layers as an intermediate cause of the changes. Future studies will concentrate on both local and integral values of the transfer parameters and on the fluid dynamics in the different structural regions. Such results will be highly important in the compilation of calculation techniques and for improved evaluation of the turbulence levels in different fluids.

The influence of the fluid physical properties may be represented by introducing into the equations the value of Pr_f with a variable exponent. For the heat transfer between a cylinder and other slender bodies and a fluid, the ex-

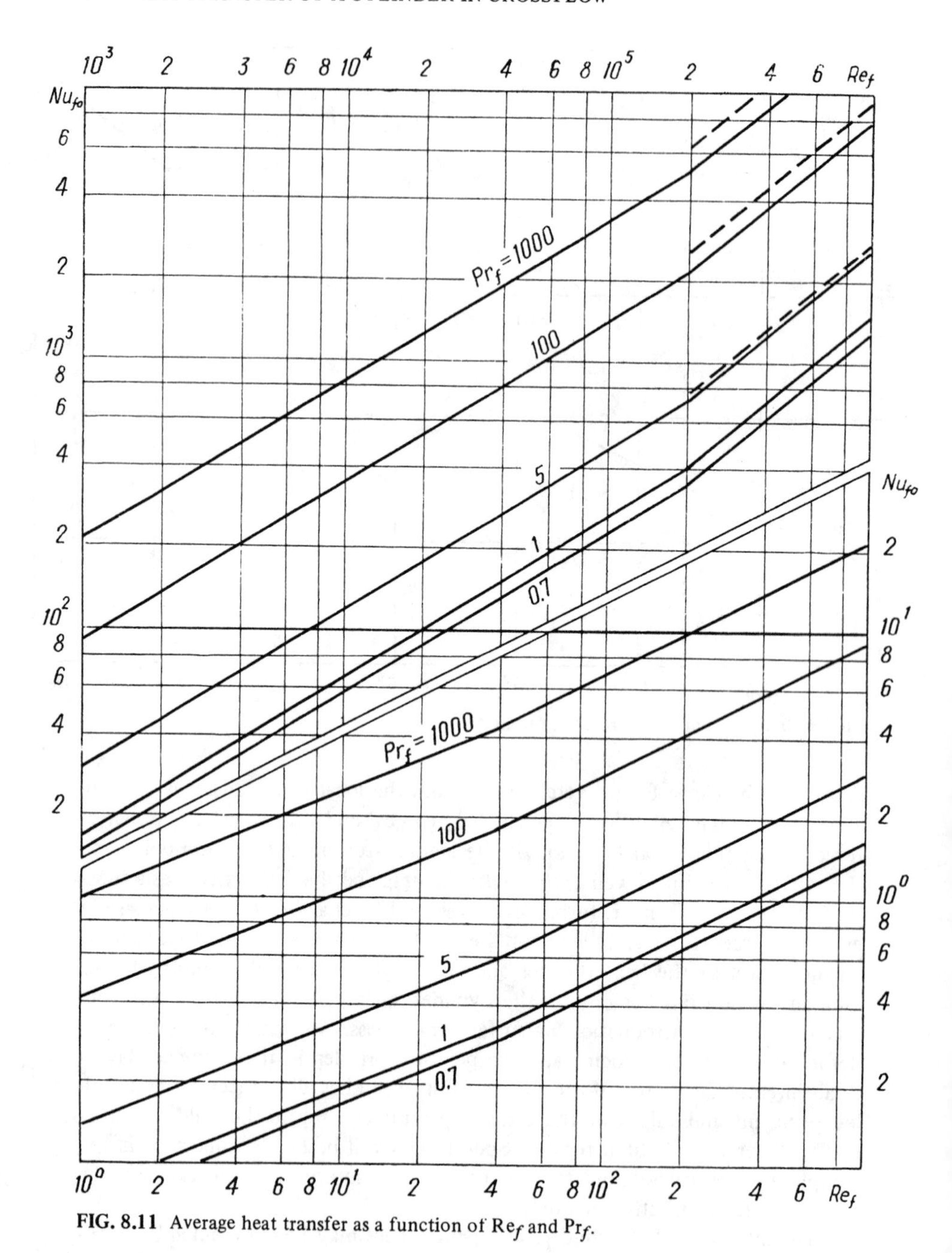

FIG. 8.11 Average heat transfer as a function of Re_f and Pr_f.

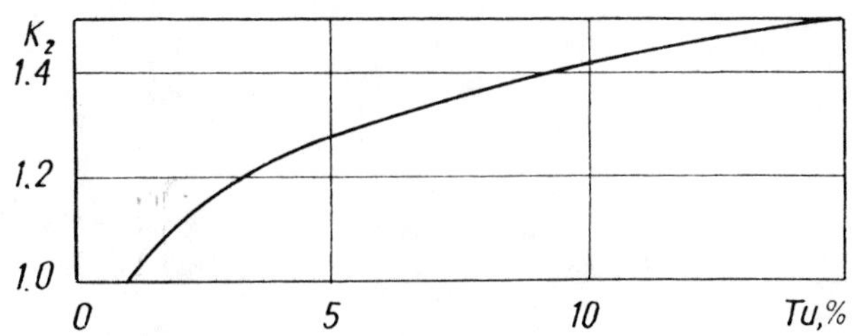

FIG. 8.12 A chart of the value of K_2 as a function of turbulence level.

ponent varies from 0.33 for the front stagnation point to a maximum of 0.45 for the rear vortical flow. Average values are 0.37 for the average heat transfer in the subcritical flow regime, and 0.4 for the critical flow regime. The effect of the temperature difference and of the heat flux direction is approximated with a sufficient accuracy by the ratio Pr_f/Pr_w to a variable exponent. The effect of the velocity is described by Re_f to a power in the range 0.4 to 0.85 for the various regions on the cylinder, with still higher values for the supercritical flow regime.

The measured heat transfer values at the front stagnation point suggest the exponent on Re_f should be 0.6, rather than the analytical value of 0.5. The phenomenon may be explained by the presence of a pseudo-laminar boundary layer and of a hypothetical cellular-vortex fluid structure. The spatial location of this structure is one object for future studies.

Our plans for future studies in the field include investigations of wider variations of the free-stream turbulence, with measurements on an extended range of fluids; we hope by this means to eliminate the remaining differences of opinion. Heat transfer studies of yawed cylinders must be extended to the critical flow regime, where both qualitative and quantitative revelations may be expected. The interaction of the laminar boundary layer and the external turbulent fluctuations is far from clear yet, as well as some other details of the fluid dynamics on cylinders in crossflow.

APPENDIX

1. PHYSICAL PROPERTIES OF FLUIDS

Table 1 Air

t, °C	ρ, kg/m³	$c_p \cdot 10^{-3}$, J/(kg·°C)	$\lambda \cdot 10^2$, W/(m·°C)	$\nu \cdot 10^6$, m²/s	Pr
20	1.205	1.005	2.59	15.06	0.703
40	1.128	1.005	2.76	16.96	0.699
60	1.060	1.005	2.90	18.97	0.696
80	1.000	1.009	3.05	21.09	0.692

Table 2 Water

20	998.2	4.18	59.9	1.006	7.02
40	992.2	4.17	63.4	0.659	4.31
60	983.2	4.18	65.9	0.478	2.98
80	971.8	4.20	67.5	0.365	2.21

Table 3 Transformer Oil (Supply 1)

20	877	1.666	11.51	27.70	350
40	865	1.788	11.28	12.10	166
60	854	1.905	11.05	6.50	96
80	842	2.026	10.82	4.30	68

Table 4 Transformer Oil (Supply 2)

20	900	1.74	11.65	25.05	344.0
40	882	1.84	11.49	10.80	153.0
60	862	1.92	11.33	5.86	85.6
80	843	2.00	11.17	3.72	55.2

Table 5 Aviation Oil

20	894.6	1.845	0.113	1091	15936
40	883.0	1.927	0.111	265	4062
60	871.3	2.008	0.110	88.6	1409

2. LOCAL HEAT TRANSFER DATA FOR CYLINDERS IN CROSSFLOW

Table 6 Circular Cylinder: Wall-to-Fluid Heat Transfer in Air

$Re_f \cdot 10^{-4}$	t_f, °C	Tu, %	α_x, $\Delta t = f(\varphi)$						
			0°	30°	60°	90°	120°	150°	180°
d=32 mm, k_q=0.16									
6.13	38.3	1.2	186 34.9	185 36.2	161 43.1	86 63.6	114 55.5	152 43.8	178 36.6
20.0	44.3	1.2	374 38.1	363 39.8	302 47.6	293 48.5	337 42.1	372 38.0	467 30.8
31.6	46.7	1.2	503 40.1	489 41.8	437 47.9	258 57.5	543 38.9	367 52.3	464 43.3
49.1	52.6	1.2	682 39.2	659 41.0	568 47.6	370 61.0	775 37.7	446 56.5	608 45.3
2.68	34.8	7.0	142 40.5	141 42.5	122 51.3	63 75.4	74 74.4	84 67.3	100 58.3
13.8	36.2	7.0	400 35.3	377 37.7	331 43.7	200 56.8	367 45.0	188 66.8	259 56.3
31.9	45.2	7.0	698 37.3	634 41.2	549 47.7	436 55.6	596 46.8	370 66.2	504 52.3
50.4	50.5	7.0	942 35.0	844 39.1	743 44.1	762 43.1	792 42.5	554 57.3	758 43.7
d=50 mm, k_q=0.25									
17.7	34.8	1.2	230 36.6	213 39.5	185 47.0	142 55.9	195 43.6	190 43.4	239 35.7
98.2	52.7	1.2	729 29.7	641 33.7	576 37.2	739 29.5	655 33.0	592 36.3	733 29.6
19.8	36.3	2.7	262 33.9	248 35.9	221 41.2	161 49.4	284 32.3	226 38.2	276 32.4
104	48.0	2.7	758 33.4	691 36.5	642 39.0	866 29.5	760 33.2	608 41.2	714 35.4
19.4	38.1	7.0	313 38.6	288 41.1	253 47.7	177 60.9	291 45.1	172 65.2	238 51.0
102	51.1	7.0	956 40.9	833 46.8	857 45.3	1192 33.2	774 50.4	718 54.6	818 47.9
13.9	34.8	15.0	369 28.8	323 33.0	279 38.1	240 43.7	244 44.4	187 54.9	245 43.7
99.7	53.0	15.0	1186 28.3	1062 31.5	1253 26.8	1358 25.0	736 45.3	686 48.7	826 40.7

Table 7 Circular Cylinder: Wall-to-Fluid Heat Transfer in Water

$Re_f \cdot 10^{-4}$	t_f, °C	Tu, %	$\alpha_x \cdot 10^{-3}$, $\Delta t = f(\varphi)$						
			0°	30°	60°	90°	120°	150°	180°
$d=30,1$ mm, $k_q=0.3$									
21.8	34.9	1.07	16.5 3.4	16.4 3.5	15,7 3.6	11,0 5.2	18,5 3,1	20.1 2,8	23,7 2.4
49,0	34,8	0,95	26,0 3,3	25.7 3.4	24,6 3.5	23.1 3.8	37,2 2,3	26.2 3.3	31,8 2,7
120	51.5	0,95	38,7 4,3	38.0 4.4	36.4 4.6	67,3 2,5	56,5 2,9	53,2 3,1	60.6 2,7
25.2	32.0	3.90	22.3 4.1	21,3 4,3	18.0 5.0	14.5 6.2	30,0 3,0	23,5 3,9	27,6 3,3
128	44.9	2,75	47.4 4,4	44,5 4.6	40.2 5.1	68,0 3,0	75,0 2.8	58,2 3.6	67,0 3,1
140	48.2	2,75	46,2 4.8	43.4 5.1	39.4 5.6	54,3 4.1	68.8 3.2	56,6 3.9	64.4 3.4
24,5	25.0	7,45	22,3 3.0	21.9 3,02	20.4 3,2	16.7 4.0	39.0 1,7	16,9 3,9	22.8 2.9
143	58.5	6.2	45.8 5.1	45.2 5.2	44.3 5.3	91,8 2.5	57.0 4,1	56,4 4,13	64.3 3,6
$d=68,4$ mm, $k_q=0.68$									
17.6	31.0	2,25	5,30 6,2	5,40 6.1	5.30 6.2	2.75 11.9	4.40 7.4	5,20 6.3	6,50 5,0
31.4	30,8	1,85	7,20 6,0	7,20 6.0	7,00 6,1	4,75 9,0	7.40 5.8	8,40 5,1	10,0 4.3
58.4	36.6	1,70	10,0 8,1	10,1 8,0	10.1 8.0	8,80 9,2	29,0 2,8	11,9 6.8	14,1 5,7
89,1	36,5	1,1	13,1 7.3	12,9 7,5	12,5 7,7	11.2 8,6	28,0 3.4	17,1 5,6	18,1 5.3
202	36,5	0,95	21,8 6,0	19.9 6,6	30,5 4.3	54,2 2.4	44,0 3,0	38,0 3.4	40,4 3.2
417	53.2	0,95	26,9 4.9	25,7 5,1	45,0 2,9	84,5 1,6	44,0 3,0	45,6 2,9	52,0 2,5
119	18,5	3.5	21,7 5,7	20,6 6,0	18,2 6,8	15,7 7,9	47,5 2,6	31,0 4,0	40,3 3,1
254	26,6	6.2	32,5 4,5	28,9 5,1	47,0 3,1	69,3 2,1	46,0 3,2	38,0 3,9	41,3 3,6

Table 7 Circular Cylinder: Wall-to-Fluid Heat Transfer in Water (*Continued*)

$Re_f \cdot 10^{-4}$	t_f, °C	Tu, %	$\alpha_x \cdot 10^{-3}$, $\Delta t = f(\varphi)$						
			0°	30°	60°	90°	120°	150°	180°
$d = 30.7$ mm, $k_q = 0.17$									
5.50	35.2	–	8.45 2.7	785 2.9	6.50 3.5	4.65 4.9	6.85 3.3	8.80 2.6	10.0 2.3
8.95	35.1	–	10.7 3.4	10.2 3.5	8.60 4.2	7.10 5.1	10.5 3.4	11.5 3.1	13.1 2.8
11.9	35.3	–	12.3 4.2	12.0 4.3	10.5 5.0	13.4 3.9	13.5 3.8	14.0 3.7	15.8 3.3
23.7	34.5	–	18.0 3.4	17.6 3.5	16.0 3.8	11.5 5.3	26.3 2.4	17.7 3.5	21.1 2.9
28.0	35.4	–	20.3 3.5	19.8 3.6	18.0 4.0	12.0 5.9	30.0 2.4	16.8 4.2	22.5 3.2
41.6	35.3	–	25.4 4.1	24.6 4.3	22.5 4.7	16.4 6.4	41.0 2.6	24.8 4.2	29.8 3.5
60.8	51.1	–	28.8 4.6	28.2 4.7	26.3 5.1	22.5 5.9	43.4 3.1	31.9 4.2	35.7 3.7
93.0	36.9	–	38.2 5.8	37.9 5.9	36.5 6.1	33.1 6.7	62.0 3.6	45.6 4.9	48.2 4.6
$d = 50.0$ mm, $k_q = 0.28$									
9.53	33.1	–	6.68 5.9	6.35 6.2	5.30 7.5	4.60 8.6	6.30 6.3	7.40 5.3	9.40 4.2
11.6	35.8	–	7.70 5.1	7.60 5.2	6.80 5.8	6.00 6.6	7.70 5.1	8.00 4.9	9.40 4.2
20.3	33.2	–	10.4 5.3	9.85 5.6	8.80 6.2	8.40 6.5	17.0 3.2	12.1 4.5	14.5 3.8
44.2	35.6	–	17.3 4.2	16.6 4.4	14.9 4.8	12.7 5.7	30.5 2.4	16.5 4.4	21.0 3.4
73.3	35.3	–	22.3 4.1	21.5 4.3	20.0 4.6	19.3 4.8	39.0 2.4	25.3 3.7	29.9 3.1
87.7	36.0	–	25.3 3.0	24.0 3.2	22.2 3.5	22.3 3.3	45.3 1.2	29.3 2.4	34.3 1.9
110	38.8	–	30.5 3.8	27.3 4.3	25.6 4.6	52.5 2.2	47.0 2.5	34.6 3.4	40.0 2.9
203	39.3	–	41.7 3.6	37.0 4.0	59.5 2.5	77.3 1.9	60.9 2.5	51.8 2.9	57.2 2.6

Table 8 Elliptic Cylinder: $d_1 = 58.3$ mm, $d_2 = 30.3$ mm

$Re_{fd} \cdot 10^{-3}$	t_f, °C	Pr_f	$\alpha_x \cdot 10^{-1}$, $\Delta t = f(\varphi)$						
			0°	15°	30°	75°	120°	165°	180°
Flow parallel to the major axis									
16.8	32.8	0.7	17.9 5.5	15.7 7.0	12.2 10.2	8.82 15.9	7.25 20.3	11.4 11.2	12.4 10.0
21.3	31.5	0.7	8.55 11.0	7.77 12.1	6.39 14.7	4.45 21.1	3.52 26.7	4.48 21.0	5.21 18.1
72.2	28.3	0.7	7.67 16.7	6.99 17.9	5.91 20.4	4.19 26.9	3.26 33.3	3.89 28.6	4.51 25.3
22.4	22.4	6.6	490 5.7	417 6.7	325 8.6	221 13.2	148 18.9	247 11.3	377 7.4
73.0	22.4	6.6	973 2.8	628 3.3	597 4.6	430 6.4	303 9.1	685 4.0	804 3.4
20.5	73.0	64	209 10.6	181 12.2	131 16.9	85.6 25.9	61.8 35.8	82.3 26.9	137 16.2
22.2	56.0	95	247 8.1	217 9.3	153 13.1	96.7 20.1	75.3 26.6	102 19.6	163 12.3
7.5	26.2	262	224 5.9	193 6.8	137 9.6	90.7 14.5	47.7 27.5	50.6 25.9	63.3 20.7
Flow parallel to the minor axis									
16.7	30.8	0.7	7.91 17.1	7.91 17.1	7.91 17.1	7.04 19.2	7.19 18.8	9.29 14.5	9.77 13.8
40.3	31.9	0.7	12.7 18.0	12.7 18.0	13.0 17.5	11.6 19.7	13.6 16.8	16.6 13.7	17.5 13.0
23.7	22.5	6.6	473 7.4	467 7.6	473 7.5	467 7.6	630 5.0	624 5.7	697 5.1
68.0	22.5	6.6	679 14.6	679 14.6	699 14.2	841 11.8	1363 7.3	920 10.8	995 10.0
20.4	73.0	64	158 23.7	157 23.9	164 22.9	158 23.7	215 17.5	241 15.6	263 14.3
3.4	56.0	95	78.9 25.3	79.6 25.1	83.2 24.0	81.7 24.4	84.1 23.7	90.6 22.0	102 19.6
4.6	26.2	263	138 22.8	138 22.8	144 21.8	141 22.3	154 20.4	167 18.8	183 17.1

Table 9 Rough-Surface Cylinder, $k_q = 0.3$

$Re_f \cdot 10^{-4}$	t_f, °C	Tu, %	$\alpha_x \cdot 10^{-3}$, $\Delta t = f(\varphi)$						
			0°	30°	60°	90°	120°	150°	180°
d=30.2 mm, k=0,15 mm									
8.7	31.4	1.7	10,8 6,4	11,6 5.6	10,8 6.4	7.5 8.7	9,1 7,1	11.2 5,8	12,0 5.4
17,0	30,5	1,25	15.2 6.3	19.6 4.9	27.4 3.5	35.7 2.6	10.2 9.4	11.6 9,0	14.4 6.7
81,0	54.6	1,0	37.0 4.9	55.0 2.9	116.0 1.4	121,0 1.3	28.5 5.6	37,0 4.4	43,5 3.7
12,0	19.7	3.2	16.0 3.8	16.7 4.5	21.2 3.5	25.0 3.0	13.0 5,8	12,9 5,8	17.4 4.3
82.0	40.7	2.4	46.5 3,5	86.5 1.9	153.5 1.0	147,0 1.1	40,0 4.0	48.0 3.4	58,5 2.8
12.0	33.9	9.5	16,0 4.8	16.9 4,6	20.0 3.8	24,0 3.2	9.0 8.5	10.2 7.5	13,0 5.9
51.0	39.5	6.5	41.0 3.3	59.0 2,3	109.0 1.2	127,0 1,1	36.0 3.7	44,0 3,0	51.0 2.6
110	46.2	6.1	57,0 3,1	81.0 2,2	126.0 1.4	143.0 1.2	58.0 3.0	57,5 3,3	65,0 2.7
d=29,1 mm, k=1,2 mm									
17,0	29.9	1,0	16.0 2.5	13,8 2.9	35.4 1.2	39.8 1,0	20,0 2.0	11,5 3,5	13.5 3.0
54.0	50,0	1.0	32.0 4,7	25.5 5.9	82.0 1.8	107.5 1.4	95,0 1,6	26.0 5.8	32,0 4.7
84,0	30,7	1,0	50,0 3.3	49.0 3.4	116.0 1.4	137.5 1.2	86.0 1.9	45.0 3.7	48.0 3.4
120	50,4	1,0	55.0 3.6	65.0 3,0	140.0 1.4	200.0 1,0	112,0 1,7	62.5 3.1	54.0 3,6
27.0	38.0	7.1	26,0 4.8	35.0 3,6	60,0 2.1	60.0 2,1	26,0 4.8	23,5 5.3	25.2 5.0
60.0	38.7	5.5	55.0 3.0	54.5 3.1	109.5 1.5	104.0 1,6	46,0 3.6	40,0 4.1	45,1 3.6
99.0	38.6	4.9	75,0 2.6	110.0 1.8	145.5 1.3	145,5 1.3	68.0 2.8	57,0 3.4	63.2 3,1
110	47,5	4.9	75,0 3,0	107,0 2.1	168,0 1,3	177,0 1.3	78.0 2.9	56.5 4.0	59,0 3,8

3. AVERAGE HEAT TRANSFER DATA FOR CYLINDERS IN CROSSFLOW

Table 10 Water Calorimeter: $d = 12$ mm, $k_q = 0.24$

t_f, °C	t_w, °C	U, m/s	Nu_f	Pr_f	$Re_f \cdot 10^{-3}$	$\frac{Nu_f}{Pr_f^{0.37}(Pr_f/Pr_w)^{0.25}}$
Wall-to-fluid heat transfer in air						
30.6	53.5	5.49	31.0	0.7	4.07	35.1
29.9	70.8	5.47	33.5	0.7	4.09	37.9
30.5	70.6	8.03	43.5	0.7	5.96	49.5
30.8	53.3	8.03	44.0	0.7	5.96	49.9
29.7	71.5	9.64	48.7	0.7	7.26	55.4
29.4	70.7	13.86	55.8	0.7	10.50	63.5
29.9	53.2	17.16	66.7	0.7	13.00	75.7
29.0	70.1	19.86	70.6	0.7	15.17	80.3
30.7	70.6	23.90	83.6	0.7	17.82	94.8
30.8	70.6	26.10	89.1	0.7	19.50	101.4
Fluid-to-wall heat transfer in air						
29.7	11.3	5.22	30.5	0.7	3.86	34.83
30.2	11.4	8.20	38.8	0.7	6.07	44.31
30.1	11.6	9.71	48.7	0.7	7.18	55.2
30.1	12.0	14.00	59.5	0.7	10.43	67.9
30.0	11.9	17.23	65.2	0.7	12.92	74.5
29.5	11.9	19.97	69.0	0.7	15.10	78.8
29.5	11.9	22.3	74.0	0.7	16.87	84.5
30.2	12.2	24.00	74.1	0.7	17.85	84.1
30.0	12.2	26.30	80.2	0.7	19.62	91.6
Wall-to-fluid heat transfer in water						
17.7	32.9	0.59	122.2	7.48	6.67	52.6
18.1	33.9	0.83	150.0	7.40	9.43	64.6
18.0	31.3	1.18	180.3	7.42	13.28	89.4
18.0	30.0	1.48	203.0	7.42	16.33	89.4
18.0	29.0	1.67	221.0	7.42	18.86	98.5
17.9	26.1	2.42	278.0	7.44	27.3	125.2
18.0	24.5	3.92	365.0	7.42	44.3	166.4
18.1	23.3	5.57	468.0	7.40	63.0	215.0
18.0	21.4	8.15	578.0	7.42	91.9	270.0

Table 10 Water Calorimeter: $d = 12$ mm, $k_q = 0.24$ (*Continued*)

t_f, °C	t_w, °C	U, m/s	Nu_f	Pr_f	$Re_f \cdot 10^{-3}$	$\frac{Nu_f}{Pr_f^{0.37}(Pr_f/Pr_w)^{0.25}}$
Fluid-to-wall heat transfer in water						
29,9	23,9	0,56	102,6	5,43	8,3	57,0
29,7	24,4	0,79	122,0	5,45	11,7	67,4
29.8	25.8	1.37	178.8	5,44	20,2	98.0
50,0	42,8	1,59	198,8	3.53	50.6	129,1
29.8	27.2	3.42	330,0	5,44	50,6	179.2
29,8	27,7	4,84	404,0	5,44	71,6	219,0
50,0	45,6	4,41	395,0	5,53	94,6	253,0
49,9	46,1	6,27	493,0	3,54	134,5	314,0
49,9	46,9	8,13	569,0	3,54	174,2	362,0
Wall-to-fluid heat transfer in transformer oil						
25,6	55,3	0,61	101,3	273	0,335	10,1
24,7	53,9	0,85	117,0	284	0,448	11,3
24,2	53,0	1,04	129,0	289	0,338	12,6
23.9	52.4	1.20	139,5	293	0,613	13,3
24.3	44.6	1,45	139,1	288	0,761	14.4
23.8	43.7	1,91	157,5	294	0,972	16,0
23,3	43,5	2,13	163,2	301	1,060	16,5
23,6	49,8	2,58	185,5	297	1,300	17,9
23,0	48,6	3,01	201,0	305	1,480	19,3
22,9	47,8	3,35	214,0	306	1,640	20,6
22,9	41,1	3,68	217,0	306	1,800	22,2
23,1	38,8	5,99	292,0	303	2,940	30,2
23,5	44,2	6,70	325,0	298	3,380	33,0
Fluid-to-wall heat transfer in transformer oil						
30,0	16,7	1,02	89,4	233	0,687	13,7
49,9	18,8	0,61	68,5	124	0,825	15,1
30,0	15,7	1,47	104,0	233	1,000	16,1
30,0	16,0	1,80	111,6	233	1,229	17,2
30,3	17,9	1,90	124,0	230	1,290	18,8
30,0	18,0	2,85	148,5	233	1,910	22,4
31,8	19,3	3,01	153,3	218	2,19	23,7
31,4	20,1	4,25	193,0	221	3,04	29,3
30,0	21,2	6,70	289,0	233	4,60	42,0
49,8	30,1	4,70	263,0	125	6,30	51,4
50,0	32,6	7,07	346,0	124	9,53	66,4

Table 11 Electric Calorimeter

t_f, °C	t_w, °C	U, m/s	Nu_f	Pr_f	$Re_f \cdot 10^{-3}$	$\frac{Nu_f}{Pr_f^{0.37}(Pr_f/Pr_w)^{0.25}}$
d=50.25 mm, k_q=0.25						
Wall-to-fluid heat transfer in air						
31.0	38.1	8.30	120	0.7	26.1	137
29.9	51.6	12.60	161	0.7	39.5	173
24.4	60.1	16.10	181	0.7	52.4	206
24.5	45.4	22.00	223	0.7	71.5	254
Wall-to-fluid heat transfer in water						
17.3	23.0	0.38	225	7.6	17.50	100
17.4	22.8	0.52	271	7.6	24.5	122
21.1	28.2	1.33	542	6.8	68.6	259
22.3	27.4	2.48	765	6.6	130.0	368
Wall-to-fluid heat transfer in transformer oil						
56.1	63.1	0.69	236	95	5.4	42
66.2	73.9	1.70	482	73	17.0	94
56.1	64.9	5.00	863	95	39.0	153
66.4	73.5	5.02	935	73	50.4	185
d=30.7 mm, k_q=0.17						
Wall-to-fluid heat transfer in water						
26.6	29.8	1.47	382	5.88	52.2	194
35.2	38.3	1.30	358	4.80	55.0	197
35.3	39.0	1.51	384	4.79	63.9	210
27.0	30.4	1.90	455	5.83	67.7	231
35.1	38.7	2.12	485	4.81	89.5	265
35.3	39.4	2.82	621	4.79	119.2	341
30.6	33.5	3.12	657	5.33	120.0	348
35.3	38.1	3.07	624	4.79	130.0	344
38.7	41.4	3.12	671	4.43	142.0	380
34.5	37.4	3.99	780	4.86	167.0	426
34.9	37.6	4.23	814	4.83	178.0	448
35.5	38.4	4.72	906	4.77	201.0	497

Table 11 Electric Calorimeter (*Continued*)

t_f, °C	t_w, °C	U, m/s	Nu_f	Pr_f	$Re_f \cdot 10^{-3}$	$\frac{Nu_f}{Pr_f^{0.37}(Pr_f/Pr_w)^{0.25}}$
35.3	38.1	4.78	896	4.79	203.0	494
34.8	38.0	5.66	928	4.84	237.0	508
34.5	37.8	5.67	980	4.85	237.0	534
35.4	38.9	6.59	995	4.78	280.0	547
36.1	39.7	6.64	1031	4.70	287.0	567
34.4	38.6	8.29	1144	4.86	346.0	620
34.8	38.8	8.29	1148	4.84	348.0	626
34.3	38.6	9.76	1278	4.87	406.0	691
35.3	39.2	9.81	1308	4.79	416.0	719
34.7	38.9	10.94	1362	4.86	458.0	739
41.5	45.7	10.92	1428	4.17	524.0	820
21.7	26.0	18.90	2020	6.69	600.0	968
51.1	55.3	10.84	1474	3.47	608.0	915
34.8	38.9	21.00	1995	4.83	880.0	1307
36.9	41.2	21.00	2520	4.60	930.0	1388

$d=50$ mm, $k_q=0.27$

Wall-to-fluid heat transfer in water

t_f, °C	t_w, °C	U, m/s	Nu_f	Pr_f	$Re_f \cdot 10^{-3}$	$\frac{Nu_f}{Pr_f^{0.37}(Pr_f/Pr_w)^{0.25}}$
33.1	39.4	1.44	507	5.03	95.3	270
35.8	41.1	1.67	588	4.74	116.0	322
33.2	38.8	2.04	627	5.02	135.0	334
35.6	40.5	2.38	807	4.75	166.0	441
33.2	37.9	3.06	934	5.02	203.0	500
35.6	40.2	3.32	1046	4.76	230.0	571
33.2	37.7	4.13	1085	5.02	274.0	582
35.5	39.8	5.17	1236	4.76	359.0	676
35.6	39.7	6.36	1412	4.76	442.0	774
35.5	39.6	7.34	1546	4.76	509.0	846
35.3	38.9	10.60	2040	4.78	733.0	1120
38.9	42.5	11.59	2190	4.43	860.0	1235
35.3	38.8	12.66	2310	4.78	877.0	1271
38.8	42.0	14.88	2910	4.43	1100.0	1642
49.6	52.9	12.69	2750	3.56	1130.0	1696
39.7	42.8	15.86	3110	4.34	1200.0	1777
39.1	41.6	24.1	4090	4.40	1800.0	2320
39.3	42.0	27.2	4420	4.38	2030.0	2520

4. AVERAGE HEAT TRANSFER FOR CYLINDERS IN TURBULENT FLOWS OF AIR AND WATER

Table 12 Air

t_f, °C	t_w, °C	U, m/s	Nu_f	Tu, %	$Re_f \cdot 10^{-3}$	$\frac{Nu_f}{Pr_f^{0.37}(Pr_f/Pr_w)^{0.25}}$	$\frac{Nu_f}{Pr_f^{0.4}(Pr_f/Pr_w)^{0.25}}$
$d=32$ mm, $k_q=0.16$							
38.3	82.7	12.4	172	1.2	61.3	196	–
41.1	87.7	12.6	234	1.2	98.9	267	–
44.3	85.5	13.3	394	1.2	200.0	450	–
50.8	93.7	13.4	557	1.2	383.0	635	642
52.6	98.7	13.4	634	1.2	491.0	723	731
36.8	79.8	13.5	253	7.0	78.0	289	–
35.4	82.5	12.9	306	7.0	106.0	349	–
38.2	88.6	13.0	445	7.0	203.0	508	–
45.2	94.0	13.8	604	7.0	319.0	689	–
50.5	93.8	13.8	837	7.0	504.0	955	965
$d=50$ mm, $k_q=0.25$							
33.5	82.5	14.2	227	1.2	95.5	259	–
34.8	78.1	13.8	360	1.2	177.0	411	–
46.6	88.8	14.9	602	1.2	433.0	687	694
49.8	90.1	15.0	785	1.2	660.0	895	905
52.7	85.5	14.9	1121	1.2	982.0	1279	1293
31.4	68.9	14.0	193	2.7	60.5	220	–
34.3	74.1	15.9	305	2.7	126.0	348	–
37.3	75.9	15.7	482	2.7	255.0	550	–
40.9	78.7	15.6	697	2.7	468.0	795	804
43.0	82.7	15.6	912	2.7	711.0	1041	1052
48.0	82.9	15.4	1243	2.7	1038.0	1419	1434
35.1	92.7	15.3	191	7.0	47.6	218	–
39.0	92.5	15.4	590	7.0	306.0	673	–
42.4	86.4	15.4	792	7.0	459.0	903	913
46.5	93.8	15.5	1092	7.0	702.0	1246	1259
51.1	96.1	15.4	1482	7.0	1026.0	1691	1709
27.8	74.1	14.8	203	15.0	47.3	232	–
35.4	76.8	15.7	373	15.0	123.0	426	–
34.4	75.5	14.4	590	15.0	237.0	673	–
38.5	86.2	14.3	885	15.0	433.0	1010	1021
48.8	80.8	15.3	1268	15.0	677.0	1446	1462
53.0	86.0	15.2	1723	15.0	997.0	1966	1987

Table 13 Water

t_f, °C	t_w, °C	U, m/s	Nu_f	Tu, %	$Re_f \cdot 10^{-3}$	$\frac{Nu_f}{Pr_f^{0.37}(Pr_f/Pr_w)^{0.25}}$	$\frac{Nu_f}{Pr_f^{0.4}(Pr_f/Pr_w)^{0.25}}$
$d=30.1$ mm, $k_q=0.3$							
34.4	40.1	1.3	335	2,05	53	180	–
34.0	39.9	2.8	536	1,45	114	281	–
35.0	38.5	4.0	665	1,25	163	364	–
35.0	38.3	5.3	816	1,07	218	447	–
34.9	38.0	6.6	1001	1,05	274	549	–
34.8	38.0	9.0	1119	0,95	371	612	–
34.8	38.0	11.9	1303	0,95	491	714	681
34.8	38.7	15.1	1692	0,95	622	921	879
34.0	38.2	21.8	2270	0,95	882	1227	1170
51.5	54.7	21.8	2400	0,95	1206	1500	1446
30.4	34.8	1.7	387	5,45	63	202	–
30.0	34.0	2.2	510	5,20	83	265	–
31.6	34.6	2.7	564	5,05	103	299	–
31.1	35.7	3.9	838	4,65	148	443	–
31.5	34.3	5.1	1013	4,25	196	541	–
30.1	34.7	7.7	1209	3,65	289	626	–
30.4	35.0	10.4	1317	3,15	390	688	–
30,6	34.8	12,7	1448	2,80	480	759	722
31.3	36.2	20.3	1981	2,70	776	1042	992
44,9	48.5	26,0	2660	2,75	1285	1578	1515
26,8	30.9	2.0	485	9,50	71	246	–
26,3	30.3	2.6	579	9,20	89	292	–
26,6	30,5	3,4	712	8,80	119	362	–
24,9	28.0	5.4	894	8,05	180	448	–
25,0	28.1	7.3	1049	7,45	245	524	–
25.2	28,1	10,1	1317	6,70	340	661	626
25,2	27,9	12.3	1577	6,40	413	971	749
25,3	27,7	17,8	2010	6,20	599	1011	958
58.3	62.8	13.8	2010	6,30	848	1295	1252
58.5	62,4	23,2	2760	6,25	1430	1785	1726
$d=50$ mm, $k_q=0.5$							
34,5	42.6	2.2	644	1.90	149	343	–
33.6	40.7	5,2	1121	1,25	349	594	–
34.1	39,7	8.9	1499	1.00	598	802	–
34,3	39.6	12.8	1807	0.95	867	972	927
34,3	38.9	18,3	2250	0.95	1239	1219	1163
46,9	51,6	27.1	3410	0.95	2320	2050	1968

Table 13 Water (*Continued*)

t_f, °C	t_w, °C	U, m/s	Nu_f	Tu, %	$Re_f \cdot 10^{-3}$	$\frac{Nu_f}{Pr_f^{0.37}(Pr_f/Pr_w)^{0.25}}$	$\frac{Nu_f}{Pr_f^{0.4}(Pr_f/Pr_w)^{0.25}}$
20.1	26.2	1.9	596	5.60	93	278	–
30.0	35.8	3.8	835	5.05	237	432	–
30.3	35.9	7.4	1339	4.20	460	695	–
30.3	35.9	9.4	1436	3.90	587	745	708
33.0	38.0	12.2	1842	3.45	807	983	937
31.8	36.3	21.8	3150	2.70	1403	1664	1584
50.0	53.2	28.0	4460	2.70	2520	2740	2640
26.1	35.1	1.5	539	10.00	85	264	–
26.1	32.5	2.4	759	9.55	139	377	–
27.9	34.6	6.0	1182	8.50	359	596	–
25.8	31.5	10.0	1609	7.50	565	803	761
25.7	30.9	12.7	1892	6.90	718	942	893
25.7	30.7	19.5	2850	6.25	1100	1420	1346
56.8	59.8	21.4	3630	6.20	2140	2340	2260

$d=68.4$ mm, $k_q=0.68$

t_f, °C	t_w, °C	U, m/s	Nu_f	Tu, %	$Re_f \cdot 10^{-3}$	$\frac{Nu_f}{Pr_f^{0.37}(Pr_f/Pr_w)^{0.25}}$	$\frac{Nu_f}{Pr_f^{0.4}(Pr_f/Pr_w)^{0.25}}$
31.0	37.6	2.0	549	2.25	176	285	–
36.6	44.0	4.3	981	1.70	416	527	–
36.6	43.3	6.0	1311	1.45	584	704	–
33.7	41.8	8.0	1529	1.25	734	794	–
36.5	42.6	9.2	1692	1.10	891	911	870
36.3	42.2	11.9	2090	1.00	1153	1120	1072
36.5	40.2	20.8	3860	0.95	2020	2100	2000
21.5	32.0	1.7	555	5.90	120	256	–
20.0	28.7	3.0	833	5.55	206	377	–
19.95	25.9	5.5	1459	5.10	371	673	–
19.85	26.3	7.2	1884	4.85	484	851	–
19.05	25.2	12.7	2290	4.10	841	1075	–
18.0	22.6	20.9	3090	3.25	1351	1348	1269
48.7	51.9	21.4	4230	3.25	2570	2570	2470
25.2	31.8	1.9	554	10.10	148	271	–
25.2	32.5	3.6	1013	9.75	272	489	–
25.2	32.1	6.4	1311	9.00	489	632	–
26.9	33.6	8.7	1565	8.60	688	769	729
26.8	34.4	11.0	1900	8.20	972	927	879
26.8	33.2	14.5	2250	7.65	1146	1105	1050
26.6	30.0	32.2	4910	6.20	2540	2370	2320
56.7	60.2	24.6	4430	6.55	3340	2810	2710

5. AVERAGE HEAT TRANSFER OF TWO CYLINDERS IN SERIES

Table 14 Water: Cylinders $d = 30$ mm, $k_q = 0.3$

t_f, °C	t_w, °C	U, m/s	Nu_f	Pr_f	$Re_f \cdot 10^{-3}$	$\frac{Nu_f}{Pr_f^{0.37}(Pr_f/Pr_w)^{0.25}}$	$\frac{Nu_f}{Pr_f^{0.4}(Pr_f/Pr_w)^{0.25}}$
Leading cylinder at $x/d = 3$							
18.9	22.7	1.6	362	7.26	45	169	–
18.8	21.7	2.5	487	7.28	72	229	–
18.7	45.0	3.8	744	7.32	110	301	–
18.5	21.6	5.0	804	7.35	144	375	–
18.9	44.0	6.5	1091	7.26	188	444	–
18.3	44.0	8.9	1244	7.40	255	500	–
18.3	45.7	13.1	1526	7.40	374	610	575
18.3	22.5	18.2	1843	7.40	520	848	799
48.7	54.6	14.6	1499	3.63	769	904	869
49.3	53.8	22.9	2180	3.58	1220	1333	1283
Rear cylinder at $x/d = 3$							
22.0	25.0	1.6	489	6.64	49	238	–
21.7	24.5	2.5	635	6.70	77	308	–
19.3	22.1	3.9	810	7.16	113	383	–
19.4	22.5	5.0	875	7.14	147	414	–
20.5	22.8	6.7	1340	6.93	201	644	608
19.8	23.0	9.1	1455	7.07	269	690	651
19.8	23.2	13.0	1804	7.06	385	855	806
22.6	26.5	18.3	2500	6.54	579	1218	1151
48.9	52.4	14.5	2530	3.61	769	1545	1490
49.1	52.1	22.6	3080	3.59	1198	1898	1827
Leading cylinder at $x/d = 6$							
20.9	25.5	1.5	371	6.85	46	176	–
20.3	25.3	2.5	502	6.96	74	237	–
20.5	24.0	3.9	715	6.94	116	340	–
20.1	23.4	5.0	871	7.00	148	414	–
20.8	24.3	6.6	964	6.87	203	460	–
21.0	25.0	9.1	1096	6.83	278	523	–
21.6	27.9	14.3	1470	6.73	441	694	656

Table 14 Water: Cylinders $d = 30$ mm, $k_q = 0.3$ (*Continued*)

t_f, °C	t_w, °C	U, m/s	Nu_f	Pr_f	$Re_f \cdot 10^{-3}$	$\frac{Nu_f}{Pr_f^{0.37}(Pr_f/Pr_w)^{0.25}}$	$\frac{Nu_f}{Pr_f^{9.4}(Pr_f/Pr_w)^{0.25}}$
21.5	27.2	17.8	1645	6.72	549	781	–
49.7	55.1	14.5	1661	3.56	780	1011	973
49.9	54.3	22.5	2190	3.54	1212	1344	1294
Rear cylinder at $x/d = 6$							
19.4	22.8	1.5	529	7.16	45	249	–
19.6	23.4	2.5	697	7.12	72	328	–
16.6	21.0	4.0	833	7.80	109	376	–
16.7	21.4	5.0	898	7.78	137	404	–
17.1	21.9	6.5	998	7.67	180	449	423
17.4	21.2	9.4	1447	7.60	262	664	625
17.4	21.2	13.8	1779	7.60	385	816	767
17.7	22.0	19.4	2290	7.54	547	1051	989
49.8	54.2	14.4	2140	3.54	772	1317	1268
49.6	53.5	22.6	2880	3.56	1209	1772	1706
Leading cylinder at $x/d = 9$							
18.9	23.7	1.5	356	7.26	45	164	–
18.7	23.7	2.7	510	7.30	77	235	–
18.9	23.5	3.9	638	7.26	112	296	–
21.2	25.9	4.7	747	6.78	144	356	–
19.3	23.1	6.6	1035	7.16	192	485	–
18.4	22.5	9.0	1098	7.37	258	508	–
18.1	22.2	13.5	1290	7.43	383	605	560
18.4	24.1	19.1	1630	7.38	546	744	701
52.1	56.9	13.2	1536	3.39	735	958	923
49.6	53.8	22.5	1927	3.56	1204	1179	1136
Rear cylinder at $x/d = 9$							
17.6	22.1	1.6	505	7.55	44	231	–
17.9	21.8	2.7	680	7.50	75	313	–
18.4	23.0	3.8	793	7.38	110	367	–
17.5	22.0	4.8	930	7.59	136	425	–
17.5	21.8	6.8	1121	7.58	191	512	482
17.6	21.8	9.5	1296	7.55	267	594	559
16.9	21.1	14.0	1631	7.72	385	741	696
17.1	21.4	19.3	2070	7.68	536	944	888
51.4	55.2	14.5	2250	3.44	799	1404	1353
51.4	54.1	22.8	3140	3.45	1250	1961	1889

6. AVERAGE HEAT TRANSFER DATA FOR ELLIPTIC CYLINDERS

Table 15 Elliptic Cylinder: $d_1 = 58.3$ mm, $d_2 = 30.3$ mm

t_f, °C	t_w, °C	U, m/s	Nu_f	Pr_f	$Re_f \cdot 10^{-3}$	$\frac{Nu_f}{Pr_f^{0.37}(Pr_f/Pr_w)^{0.25}}$
Flow parallel to the major axis						
32.8	55.2	4.60	91	0.7	16.8	102
31.5	51.0	5.90	104	0.7	21.3	117
32.0	54.2	7.90	128	0.7	28.5	142
31.8	52.8	9.15	137	0.7	33.1	154
28.3	45.4	10.70	230	0.7	72.2	258
22.4	28.6	0.18	136	6.6	11.0	67
22.4	34.6	0.37	221	6.6	22.4	104
22.4	30.0	0.63	354	6.6	39.0	173
22.4	27.5	1.20	526	6.6	73.0	265
22.4	27.8	1.79	635	6.6	109.3	314
22.4	34.4	2.39	800	6.6	146.0	388
56.0	77.0	0.61	229	95	5.5	41
26.2	43.9	2.41	370	262	7.5	45
73.0	98.2	0.61	249	64	8.2	52
73.0	97.5	1.49	470	64	20.0	97
56.0	73.8	2.47	577	95	22.2	105
56.0	78.5	2.68	636	95	23.8	113
73.0	92.6	2.42	631	64	32.8	134
Flow parallel to the minor axis						
30.8	48.0	8.90	89	0.7	16.7	101
26.9	44.7	12.80	114	0.7	24.7	129
33.3	48.1	18.20	136	0.7	33.7	154
31.9	48.8	21.50	152	0.7	40.3	172
22.4	29.0	0.75	268	6.6	23.7	132
22.4	34.2	1.41	422	6.6	44.7	195
22.4	25.8	2.15	534	6.6	68.0	265
26.2	47.2	1.77	287	262	2.8	30
56.0	80.3	0.74	219	95	3.4	36
73.0	104.6	0.72	243	64	5.0	46
56.0	81.3	1.80	361	95	8.4	58
56.0	77.1	2.95	469	95	13.7	76
73.0	92.8	2.91	514	64	20.4	101

7. EXPERIMENTAL DATA ON THE FLUID DYNAMICS OF CIRCULAR CYLINDERS

Table 16 Pressure Distribution in Air

$Re_f \cdot 10^{-4}$	t_f, °C	Tu, %	$\overline{p}$, $\Delta p \cdot 10^{-3} = f(\varphi)$						
			0°	30°	60°	90°	120°	150°	180°
$d = 50$ mm, $k_q = 0,25$									
6,08	29.7	1.2	1.00	0.23	−1.45	−1.54	−1.56	−1.62	−1.64
			0.12	0.03	−0.18	−0.19	−0.19	−0.20	−0.20
33.0	44.6	1.2	1.00	0.26	−1.34	−1.88	−0.46	−0.38	−0.36
			0.73	0.19	−0.98	−1.37	−0.33	−0.28	−0.26
73.2	53.0	1.2	1.00	0.36	−1.14	−1.54	−0.51	−0.53	−0.48
			1.63	0.59	−1.86	−2.51	−0.83	−0.87	−0.78
8.64	32.4	2.7	1.00	0.28	−1.30	−1.20	−1.19	−1.22	−1.20
			0.24	0.06	−0.31	−0.28	−0.28	−0.29	−0.28
7.55	26.9	2.7	1.00	0.20	−2.08	−1.81	−1.75	−1.88	−1.88
			0.15	0.03	−0.31	−0.27	−0.26	−0.28	−0.28
18.8	25.7	2.7	1.00	−0.14	−2.31	−3.16	−1.15	−0.86	−0.84
			0.38	−0.05	−0.87	−1.19	−0.43	−0.32	−0.31
27.8	27.2	2.7	1.00	−0.06	−1.96	−2.76	−0.96	−0.65	−0.65
			0.56	−0.03	−1.11	−1.55	−0.54	−0.36	−0.36
34.4	29.5	2.7	1.00	−0.40	−2.17	−2.58	−0.71	−0.64	−0.62
			0.70	−0.28	−1.53	−1.81	−0.50	−0.45	−0.44
77.2	33.9	2.7	1.00	0.00	−1.32	−2.38	−0.93	−0.91	−0.88
			1.46	2.93	−2.81	−3.49	−1.36	−1.33	−1.28
6.6	31.6	7.0	1.00	0.06	−1.81	−2.28	−1.00	−0.85	−0.77
			0.15	0.01	−0.26	−0.33	−0.15	−0.12	−0.11
35.5	38.2	7.0	1.00	0.15	−1.92	−2.80	−0.94	−0.78	−0.72
			0.79	0.12	−1.51	−2.20	−0.74	−0.61	−0.57
78.7	45.7	7.0	1.00	0.14	−1.82	−2.45	−0.38	−0.37	−0.91
			1.79	0.08	−3.26	−4.38	−1.75	−1.73	−1.63
4.8	34.8	15.0	1.00	0.27	−1.84	−2.65	−1.25	−0.97	−0.92
			0.10	0.03	−0.19	−0.27	−0.13	−0.10	−0.10
9.0	30.8	15.0	1.00	0.31	−1.80	−2.72	−1.00	−0.77	−0.72
			0.19	0.06	−0.34	−0.51	−0.19	−0.14	−0.13
33.9	45.0	15.0	1.00	0.40	−1.78	−2.97	−1.12	−0.93	−0.88
			0.77	0.31	−1.36	−2.28	−0.86	−0.72	−0.88
51.7	46.4	15.0	1.00	0.02	−2.27	−3.07	−1.24	−1.19	−1.12
			1.17	0.02	−2.66	−3.60	−1.45	−1.40	−1.32
75.4	52.2	15.0	1.00	0.20	−1.81	−2.54	−0.93	−0.89	−0.83
			1.72	0.35	−3.11	−4.38	−1.61	−1.53	−1.44

Table 17 Pressure Distribution in Water

$Re_f \cdot 10^{-4}$	t_f, °C	Tu, %	$\bar{p}$, $\Delta p \cdot 10^{-4} = f(\varphi)$						
			0°	30°	60°	90°	120°	150°	180°
d=30,1 mm, k_q=0,3									
7,92	35,0	1,4	1,00	−0,01	−1,83	−1,83	−1,88	−1,97	−2,03
			0,18	−0.00	−0,33	−0.33	−0,34	−0.36	−0,37
10,2	29,6	1,2	1,00	−0,03	−1,98	−1,91	−1,90	−1,96	−1,93
			0,38	−0,01	−0,76	−0,73	−0,72	−0,75	−0,74
15,6	30,5	1,0	1.00	−0,01	−1,81	−1,81	−1,67	−1,67	−1,69
			0,84	−0,01	−1,54	−1,54	−1,42	−1,42	−1,44
20,1	31,3	1,0	1,00	−0,03	−2,00	−2,51	−1,44	−1,39	−1,34
			1,47	−0,04	−2,75	−3,45	−1,98	−1,90	−1,84
30,8	32,1	0,95	1,00	−0,13	−2,06	−3,03	−1,08	−0,63	−0,60
			3,10	−0,40	−6,45	−9,47	−3,37	−1,96	−1,88
43,4	32,3	0,95	1,00	−0,10	−1,92	−2,76	−0,92	−0,63	−0,60
			6,01	−0,61	−11,81	−16,95	−5,65	−3,87	−3,69
70,6	51,6	0,95	1,00	0,11	−1,82	−2,88	−0,91	−0,62	−0,59
			7,65	0,88	−14,71	−23,30	−7,36	−5,00	−4,81
93,8	56,7	0,95	1,00	0,16	−1,72	−2,80	−0,99	−0,69	−0,66
			11,47	1,96	−20,90	−33,90	−11,96	−8,34	−7,94
3,65	34,8	10,0	1,00	−0,15	−1,76	−1,66	−1,68	−1,76	−1,88
			0,04	−0,01	−0,07	−0,06	−0,07	−0,07	−0,07
5,24	35,0	9,5	1,00	−0,20	−1,91	−1,76	−1,73	−1,73	−1,75
			0,08	−0,02	−0,15	−0,14	−0,14	−0,14	−0,14
13,6	30,6	8,2	1,00	−0,28	−2,37	−2,60	−0,93	−0,86	−0,82
			0,63	−0,18	−1,54	−1,68	−0,60	−0,56	−0,53
18,7	30,8	7,5	1,00	−0,45	−2,68	−3,04	−0,83	−0,82	−0,77
			1,20	−0,55	−3,26	−3,69	−1,01	−0,99	−0,93
31,0	30,9	6,5	1,00	−0,26	−2,37	−2,55	−0,77	−0,76	−0,73
			3,29	−0,86	−7,89	−8,48	−2,56	−2,53	−2,41
47,0	34,6	6,2	1,00	−0,20	−2,00	−2,23	−0,74	−0,67	−0,62
			5,86	−1,17	−11,84	−13,20	−4,37	−3,96	−3,66
61,5	34,9	6,2	1,00	−0,29	−2,54	−3,10	−1,27	−1,14	−1,03
			11,57	−3,24	−28,10	−34,30	−14,02	−12,65	−11,38
71,8	39,4	6,2	1,00	−0,45	−2,39	−2,36	−1,00	−0,91	−0,95
			11,96	−5,69	−30,20	−29,70	−12,55	−11,47	−11,96

Table 17 Pressure Distribution in Water (*Continued*)

$Re_f \cdot 10^{-4}$	t_f, °C	Tu, %	$\bar{p}$, $\Delta p \cdot 10^{-4} = f(\varphi)$						
			0°	30°	60°	90°	120°	150°	180°
$d=50$ mm, $k_q=0.5$									
19.6	31.3	1.2	1.00 0.47	−0.09 −0.04	−2.60 −1.22	−2.34 −1.09	−2.11 −0.99	−2.03 −0.95	−1.97 −0.92
25.4	32.8	1.0	1.00 0.75	−0.30 −0.23	−3.16 −0.24	−5.13 −3.93	−2.81 −2.15	−1.59 −1.22	−1.56 −1.20
33.9	32.4	0.95	1.00 1.34	−0.43 −0.57	−3.27 −4.38	−5.57 −7.47	−3.47 −4.65	−1.65 −2.21	−1.66 −2.22
41.1	32.6	0.95	1.00 1.93	−0.58 −1.13	−3.58 −7.01	−6.17 −12.10	−4.05 −7.93	−2.01 −3.93	−2.09 −4.10
48.6	32.6	0.95	1.00 2.71	−0.56 −1.53	−3.52 −9.64	−6.09 −16.70	−3.68 −10.08	−2.28 −6.25	−2.22 −6.09
63.0	33.2	0.95	1.00 4.87	−0.47 −0.21	−3.28 −14.71	−6.16 −27.70	−4.70 −21.10	−2.43 −10.89	−2.14 −9.61
80.8	33.8	0.95	1.00 6.77	−0.96 −0.69	−2.92 −21.00	−6.01 −43.10	−4.66 −33.40	−2.58 −18.53	−2.17 −15.59
99.0	35.1	0.95	1.00 9.61	−0.38 −0.39	−2.84 −29.00	−6.10 −62.40	−4.83 −49.40	−2.67 −27.40	−2.21 −22.70
$d=70$ mm, $k_q=0.7$									
9.80	35.8	2.0	1.00 0.05	−0.90 −0.04	−6.55 −0.33	−9.66 −0.48	−9.32 −0.47	−8.62 −0.43	−8.60 −0.43
13.7	36.2	1.75	1.00 0.10	−0.84 −0.08	−6.32 −0.60	−9.35 −0.89	−8.25 −0.78	−8.25 −0.78	−8.25 −0.78
25.0	35.3	1.25	1.00 0.33	−0.99 −0.33	−6.77 −2.26	−12.81 −4.27	−5.52 −1.84	−4.29 −1.43	−4.31 −1.44
34.0	35.2	1.10	1.00 0.61	−0.98 −0.61	−6.44 −3.94	−12.59 −7.71	−5.51 −3.38	−4.27 −2.61	−4.27 −2.61
42.0	35.2	1.00	1.00 0.93	−0.92 −0.86	−6.48 −6.08	−12.91 −12.10	−7.82 −7.33	−4.56 −4.27	−4.48 −4.20
48.0	35.2	1.00	1.00 1.24	−0.92 −1.14	−6.40 −7.97	−12.65 −15.75	−7.19 −8.95	−4.67 −5.82	−4.58 −5.70
85.8	35.7	0.95	1.00 3.82	−1.08 −4.12	−7.61 −29.12	−15.65 −53.90	−7.04 −27.00	−6.89 −26.40	−6.30 −24.10
93.8	35.0	0.95	1.00 4.51	−0.91 −4.32	−7.75 −36.70	−16.78 −79.40	−7.92 −37.50	−7.42 −35.10	−6.76 −32.00

8. DISTRIBUTION OF THE SHEAR STRESS AND PRESSURE

Table 18 Circular Cylinder: $d = 50$ mm, $k_q = 0.25$

$Re_f \cdot 10^{-4}$	t_f, °C	Tu, %	$\bar{\tau}$, $\bar{p} = f(\varphi)$ 0°	30°	60°	90°	120°	150°	180°
Air									
1.86	29.3	0.5	0	3.30	6.05	−0.34	−0.15	−0.25	0
			1.00	−0.19	−1.81	−2.00	−2.05	−2.24	−2.48
1.86	29.3	3.5	0	3.40	6.40	−0.30	−0.30	−0.25	0
			1.00	−0.17	−1.90	−2.05	−2.00	−2.19	−2.33
1.86	29.3	7.0	0	3.62	7.35	−0.30	−1.00	−0.59	0
			1.00	−0.14	−1.90	−1.93	−1.85	−1.95	−2.05
2.76	29.4	0.5	0	3.20	6.50	−0.30	−0.20	−0.24	0
			1.00	−0.23	−2.01	−1.92	−1.96	−2.16	−2.40
2.76	29.4	3.5	0	3.60	6.80	−0.10	−0.20	−0.34	0
			1.00	−0.25	−2.07	−2.06	−2.09	−2.21	−2.35
2.76	29.4	7.0	0	4.10	7.80	0.50	−0.20	−0.18	0
			1.00	−0.21	−2.20	−2.06	−1.83	−1.89	−1.98
4.20	29.5	0.5	0	3.75	5.40	−0.15	−0.08	−0.12	0
			1.00	−0.15	−1.84	−1.77	−1.82	−1.96	−2.20
4.20	29.5	7.0	0	4.20	6.45	0.56	−0.12	−0.07	0
			1.00	−0.12	−2.00	−1.95	−1.39	−1.36	−1.38
Water									
2.95	20.2	0.76	0	4.10	6.60	0.12	0	0.24	0
			1.00	−0.13	−1.93	−2.00	−1.98	−2.18	−2.21
2.95	20.2	3.5	0	4.85	7.10	0.49	0	−0.12	0
			1.00	−0.07	−2.04	−2.05	−1.95	−2.07	−2.12
2.95	20.2	9.9	0	5.25	7.55	0.74	0	−0.25	0
			1.00	−0.10	−2.10	−2.10	−1.87	−1.93	−1.98
5.12	21.3	0.92	0	3.55	5.80	0.07	−0.05	−0.29	0
			1.00	−0.03	−1.87	−1.90	−1.91	−2.11	−2.32
5.12	21.3	3.5	0	3.80	6.40	0.43	−0.05	−0.22	0
			1.00	−0.16	−2.16	−2.23	−2.87	−1.77	−1.71
5.12	21.3	9.9	0	4.00	6.85	0.94	−0.05	−0.07	0
			1.00	−0.24	−2.38	−2.38	−1.71	−1.67	−1.65
9.74	22.3	1.24	0	3.00	6.20	0	−0.30	−0.47	0
			1.00	−0.03	−1.91	−1.82	−1.82	−1.86	−1.85
9.74	22.3	3.5	0	3.30	6.50	1.88	−0.10	−0.16	0
			1.00	−0.14	−2.36	−3.03	−1.40	−1.40	−1.40

Table 18 Circular Cylinder: $d = 50$ mm, $k_q = 0.25$ (*Continued*)

$Re_f \cdot 10^{-4}$	t_f, °C	Tu, %	$\bar{\tau}$, $\bar{p} = f(\varphi)$						
			0°	30°	60°	90°	120°	150°	180°
9,74	22,3	9,9	0	3.60	6,70	3.42	0.10	−0,03	0
			1.00	−0,12	−2.50	−3.38	−1.21	−1.20	−1.19
13.88	32.3	1.4	0	4,82	5.77	0	−0.24	−0.24	0
			1.00	−0.34	−2,06	−1.81	−1.72	−1.72	−1.72
13.88	32.3	9.1	0	4.34	7.57	3.13	0.36	−0.24	0
			1.00	−0.18	−2.40	−3.81	−2.40	−1.33	−1.33
44.5	31.2	0.9	0	2.88	6.73	5.10	4.13	−0.01	0
			1.00	−0.35	−3.04	−4.22	−2.25	−1.57	−1.55
44.5	31.2	6.75	0	2.97	6.97	7.29	2.56	−0.08	0
			1,00	−0.36	−3.04	−4.17	−2,25	−1.53	−1.54
Transformer oil									
0.37	26.7	0.3	0	4.45	5.75	−0.11	0.05	0.11	0
			1.00	−0.02	−1.91	−1.79	−1.68	−1.66	−1.68
0.37	26.7	0.8	0	4.49	5.80	−0.21	−0.23	−0.21	0
			1.00	−0.02	−1.95	−1.88	−1.74	−1.73	−1.75
0.37	26.7	3.8	0	4.60	6.02	−0.11	0.05	−0.04	0
			1.00	−0.08	−2.01	−1.96	−1.85	−1.89	−1.95
0.37	26.7	8.1	0	4.80	6.40	−0.04	0.10	0	0
			1,00	−0.03	−2.11	−2.13	−2.02	−2.08	−2.15
0.61	26.3	0.3	0	3.60	5.20	−0.06	0.05	0.03	0
			1.00	−0.00	−1.98	−1.80	−1.81	−1.86	−1.96
0.61	26.3	0.8	0	3.70	5.35	−0.16	0.05	0	0
			1.00	−0.04	−0.21	−1.96	−1.86	−1.98	−2.07
0.61	26.3	3.8	0	3.85	5.62	−0.13	0.04	−0.06	0
			1.00	0.04	−0.21	−2.10	−2.01	−2.15	−2.26
0.61	26.3	8.1	0	4.02	5.96	0.03	0.13	−0.03	0
			1.00	−0.02	−2.23	−2.17	−2.09	−2.22	−2.36
0.98	27.2	0.3	0	3.45	5.55	−0.02	0.08	0.05	0
			1.00	−0.13	−1.94	−1.95	−2.02	−2.05	−2.23
0.98	27.2	0.8	0	3.55	5.62	−0.02	0.09	0.01	0
			1.00	−0.13	−2.08	−2.08	−2.03	−2.18	−2.34
0.98	27.2	3.8	0	3.68	5.75	−0.02	0.08	−0.07	0
			1.00	−0.14	−2.13	−2.14	−2.07	−2.20	−2.34
0.98	27.2	8.1	0	3.95	6.12	0.03	0.10	−0.09	0
			1.00	0.17	−2.27	−2.18	−2.14	−2.30	−2.50

Table 19 Elliptic Cylinder: $d_1 = 60$ mm, $d_2 = 30$ mm

			$\bar{\tau}$, $\bar{p} = f(\varphi)$						
$Re_f \cdot 10^{-4}$	t_f, °C	Tu, %	0°	30°	60°	90°	120°	150°	180°
Flow of air parallel to the major axis									
2.24	29.3	0.5	0	3.50	0.70	0.16	−0.70	−0.55	0
			1.00	−1.00	−1.20	−1.05	−0.87	−0.95	−1.00
2.24	29.3	3.5	0	4.20	0.80	0.16	−0.80	−0.75	0
			1.00	−1.05	−1.26	−1.09	−0.94	−0.95	−0.86
2.24	29.3	5.5	0	4.40	0.85	0.16	−0.85	−0.75	0
			1.00	−1.14	−1.32	−1.14	−0.98	−0.87	−0.76
2.24	29.3	7.0	0	4.60	1.30	0.63	−0.50	−0.55	0
			1.00	−1.24	−1.40	−1.24	−1.06	−0.81	−0.71
5.08	30.0	0.5	0	3.70	1.10	0.06	−0.05	−0.15	0
			1.00	−0.85	−1.02	−0.74	−0.78	−0.74	−0.69
5.08	30.0	3.5	0	4.20	2.30	1.08	0	−0.30	0
			1.00	−1.01	−1.28	−1.15	−0.93	−0.71	−0.53
5.08	30.0	5.5	0	4.50	2.60	1.45	0.30	−0.20	0
			1.00	−1.14	−1.38	−1.28	−1.06	−0.70	−0.44
5.08	30.0	7.0	0	4.80	2.85	1.99	0.55	−0.20	0
			1.00	−1.24	−1.50	−1.41	−1.16	−0.65	−0.41
Flow of water									
4.17	27.5	0.8	0	3.55	1.60	0.13	−0.10	−0.02	0
			1.00	−0.90	−1.03	−0.81	−0.80	−0.81	−0.67
4.11	27.5	9.9	0	4.65	2.95	1.35	0.20	−0.20	0
			1.00	−1.41	−1.58	−1.38	−1.21	−0.87	−0.61
7.07	27.5	0.9	0	4.25	1.90	0.08	−0.25	−0.25	0
			1.00	−0.92	−1.20	−1.04	−0.95	−0.72	−0.64
7.07	27.5	9.0	0	5.05	2.75	1.26	0.25	−0.15	0
			1.00	−1.43	−1.70	−1.58	−1.43	−0.90	−0.41
Flow of transformer oil									
0.38	23.5	0.3	0	4.90	1.50	0.31	−0.15	−0.40	0
			1.00	−0.98	−1.12	−0.94	−0.86	−0.79	−0.81
0.38	23.5	8.1	0	5.65	1.85	0.50	−0.10	−0.50	0
			1.00	−1.15	−1.24	−1.10	−1.00	−0.91	−0.91
0.64	23.6	0.3	0	4.15	1.10	0.28	−0.20	−0.20	0
			1.00	−0.98	−1.06	−0.91	−0.82	−0.79	−0.79
0.64	23.6	8.1	0	4.75	1.65	0.37	−0.06	−0.15	0
			1.00	−1.07	−1.18	−1.05	−0.92	−0.87	−0.86

Table 19 Elliptic Cylinder: $d_1 = 60$ mm, $d_2 = 30$ mm (*Continued*)

$Re_f \cdot 10^{-4}$	t_f, °C	*Tu*, %	$\bar{\tau}$, $\bar{p} = f(\varphi)$ 0°	30°	60°	90°	120°	150°	180°
Flow of air parallel to the minor axis									
1.12	29,1	0,5	0 1.00	0.20 0,76	4.41 −0.48	0 −3,00	−0,10 −3.24	−0.20 −3.58	0 −3.76
1.12	29,1	3,5	0 1.00	0.20 0.76	4.67 −0,48	0 −3,05	−0,15 −3,29	−0.20 −3.58	0 −3.67
1.12	29,1	5,5	0 1,00	0,20 0,76	4.83 −0.48	0 −3.14	−0,15 −3.33	−0.25 −3.57	0 −3.62
1.12	29.1	7.0	0 1.00	0.25 0.76	5.62 −0.48	0 −3,19	−0.15 −3,38	−0.20 −3.56	0 −3.52
2,32	30.2	0,5	0 1.00	0.40 0.74	4.60 −0.44	0.04 −2.92	−0,15 −3,27	−1,00 −3.65	0 −3.82
2,32	30.2	3.5	0 1.00	0.45 0.74	6.62 −0,49	0 −2,98	−0.30 −3,36	−1.25 −3,70	−0.01 −3.66
2.32	30.2	5,5	0 1.00	0.47 0.74	7,52 −0,51	−0.05 −3.11	−0,35 −3.46	−1,30 −3.66	0 −3,61
2,32	30.2	7.0	0 1,00	0.50 0.74	8,85 −0,46	−0,08 −3.20	−0.45 −3,53	−1.40 --3.66	0 −3,49
Flow of water									
2,08	27.5	0.8	0 1.00	0.70 0.82	4.55 −0,42	−0.10 −2,85	−0.30 −3.28	−1.90 −3,73	0 −3,73
2.08	27,5	9.9	0 1.00	0.80 0.82	5.50 −0.47	0.09 −3.16	−0,70 −3.62	−1,50 −3.70	0 −3.53
3.53	27,5	0.9	0 1.00	0.45 0.84	4,60 −0.47	−0.18 −3,00	−0,30 −3.48	−1.10 −3.86	0 −3.90
3.53	27.5	9.9	0 1.00	0.45 0.84	6.20 −0.41	−0.18 −3.27	−0.30 −3.68	−1.10 −3.86	0 −3.70
Flow of transformer oil									
0.19	23.7	0.3	0 1.00	0.70 0,71	4.20 −0.44	−0.25 −2.89	0.15 −2,98	−0.30 −3.11	0 −3.11
0,19	23.7	8.1	0 1,00	0.90 0.71	5.72 −0.55	−0.21 −3.00	0.20 −3.12	−0.35 −3.22	0 −3.16
0.32	23.7	0,3	0 1.00	0.60 0.80	4.05 −0.37	−0.17 −2.78	0.02 −2.99	−0.30 −3.13	0 −3.16
0.32	23.7	8.1	0 1.00	0,65 0,80	4.95 −0.53	−0.15 −2.94	0 −3.17	−0.40 −3.27	0 −3,28

9. EXPERIMENTAL DATA ON THE HYDRAULIC DRAG

Table 20 Pressure Drag of Circular Cylinders

$Re_f \cdot 10^{-4}$	U_∞, m/s	t_f, °C	Pr_f	Tu, %	C_w
Air, $d = 50$ mm, $k_q = 0.25$					
6.08	10.8	29.7	0.7	1.2	1.66
33.00	11.3	44.6	0.7	1.2	0.49
73.20	11.1	53.0	0.7	1.2	0.69
7.55	10.5	26.9	0.7	2.7	1.30
18.80	10.9	25.7	0.7	2.7	0.42
77.20	9.9	33.9	0.7	2.7	0.47
6.60	11.8	31.6	0.7	7.0	0.79
35.5	11.6	38.2	0.7	7.0	0.72
78.7	11.5	45.7	0.7	7.0	0.86
4.80	11.4	34.8	0.7	15.0	1.01
33.90	11.6	45.0	0.7	15.0	1.05
75.30	11.4	52.2	0.7	15.0	0.83
Water, $d = 30.1$ mm, $k_q = 0.3$					
7.92	1.9	35.0	4.82	1.4	1.70
20.10	5.3	31.3	5.24	1.0	1.06
43.40	11.1	32.3	5.12	0.95	0.44
93.80	15.7	56.7	3.15	0.95	0.52
3.65	0.9	34.8	4.84	10.0	1.46
13.60	3.6	30.6	5.34	8.2	0.33
47.00	10.9	34.6	4.86	6.2	0.29
71.80	15.9	39.4	4.38	6.2	0.29
d=50mm, k_q=0.5					
19.60	3.1	31.3	5.24	1.2	1.37
33.90	5.2	32.4	5.12	0.95	1.34
48.60	7.4	32.6	5.08	0.95	1.66
80.80	12.0	33.8	4.95	0.95	2.31
99.00	14.3	35.1	4.81	0.95	2.48
d=70mm, k_q=0.7					
9.80	1.0	35.8	4.75	2.0	6.53
34.00	3.5	35.2	4.80	1.1	2.77
48.00	5.0	35.2	4.80	1.0	3.53
93.80	9.8	35.0	4.82	0.95	4.89

Table 21 Hydraulic Drag Coefficients of Circular Cylinders: $d = 50$ mm, $k_q = 0.25$

U, m/s	$Re_f \cdot 10^{-4}$	Re_t	C_f	C_w	C_D
1.56	0.448	13.44	0.042	0.910	0.952
1.57	0.452	36.20	0.041	0.908	0.949
2.60	0.736	22.08	0.030	1.037	1.067
2.64	0.747	59.70	0.030	1.035	1.065
2.67	0.758	288.0	0.029	1.114	1.143
2.71	0.768	622.0	0.030	1.072	1.102
4.00	1.217	36.5	0.023	1.130	1.153
4.08	1.243	99.5	0.023	1.198	1.221
4.13	1.250	475.0	0.023	1.140	1.163
4.18	1.270	1030.0	0.024	1.100	1.124
7.43	2.330	116.5	0.016	1.203	1.219
7.50	2.350	822.5	0.017	1.104	1.121
7.57	2.380	1310.0	0.021	1.030	1.051
7.63	2.393	1675.0	0.017	0.943	0.960
10.98	3.440	172.0	0.013	1.170	1.183
11.07	3.470	1214.0	0.014	1.163	1.177
11.20	3.510	1930.0	0.016	0.978	0.994
11.30	3.545	2480.0	0.015	0.890	0.905
16.20	5.070	253.0	0.011	1.174	1.185
16.40	5.135	1800.0	0.012	1.082	1.094
16.60	5.200	2860.0	0.013	0.818	0.831
16.74	5.240	3670.0	0.013	0.676	0.689
0.76	3.780	3740.0	0.015	0.751	0.766
1.22	6.235	567.0	0.010	1.217	1.227
1.28	6.550	2290.0	0.011	0.834	0.845
1.32	6.760	6690.0	0.012	0.543	0.555
2.25	11.800	1463.0	0.008	1.080	1.088
2.47	12.930	4525.0	0.008	0.536	0.544
2.51	13.200	13 070.0	0.009	0.431	0.440
2.76	17.570	2460.0	0.007	0.798	0.805
3.00	18.600	16 900.0	0.008	0.401	0.409
6.00	39.300	3930.0	0.006	0.528	0.534
5.90	38.600	3126.0	0.007	0.428	0.435
10.30	65.500	5900.0	0.005	0.487	0.492
10.30	65.500	44 200.0	0.006	0.475	0.461

Table 22 Hydraulic Drag Coefficients of Elliptic Cylinders: $d_1 = 60$ mm, $d_2 = 30$ mm

U, m/s	$Re_f \cdot 10^{-4}$	Re_t	C_f	C_w	C_D
Flow parallel to the major axis					
2.69	0.808	24.24	0.038	0.601	0.639
2.78	0.835	676.35	0.042	0.568	0.610
1.61	0.482	14.46	0.057	0.580	0.637
1.67	0.499	404.19	0.059	0.563	0.622
16.92	6.350	317.50	0.012	0.503	0.515
17.97	6.740	2359.0	0.014	0.329	0.343
18.37	6.890	3789.5	0.015	0.225	0.240
18.78	7.040	4928.0	0.016	0.184	0.200
7.74	2.914	145.7	0.018	0.585	0.603
7.83	2.947	1031.4	0.020	0.501	0.521
7.93	2.985	1641.7	0.014	0.431	0.445
8.08	3.200	2240.0	0.021	0.309	0.330
1.30	9.140	840.9	0.011	0.470	0.481
1.43	10.100	9999.0	0.015	0.125	0.140
0.74	5.225	397.1	0.013	0.535	0.548
0.79	5.565	1947.7	0.016	0.418	0.434
0.83	5.860	5801.4	0.018	0.192	0.210
Flow parallel to the minor axis					
2.69	0.404	12.12	0.023	1.594	1.617
2.79	0.418	338.98	0.024	1.503	1.527
1.64	0.246	7.39	0.030	1.460	1.490
1.68	0.252	95.76	0.030	1.420	1.450
1.69	0.254	205.74	0.032	1.400	1.432
16.08	3.010	150.50	0.008	1.676	1.684
16.56	3.100	1085.0	0.008	1.575	1.583
16.54	3.099	1704.4	0.090	1.579	1.669
16.70	3.125	2187.5	0.009	1.533	1.542
7.74	1.460	511.0	0.012	1.620	1.632
7.80	1.470	808.5	0.012	1.572	1.584
7.83	1.476	1033.2	0.013	1.565	1.578
1.28	4.500	414.0	0.006	1.806	1.812
1.29	4.570	1599.5	0.006	1.740	1.746
1.33	4.685	4638.1	0.006	1.677	1.683
0.75	2.680	203.68	0.009	1.72	1.729
0.76	2.695	943.25	0.008	1.70	1.708
0.79	2.790	2762.1	0.008	1.603	1.611

REFERENCES

1. Kenelly, A. E., Wright, C. A., and van Bylevet, I. S. 1909. The convection of heat from small copper wires. *Trans. Am. Inst. Elec. Eng.* 28:363–393.
2. King, L. W. 1914. On the convection of heat from small cylinder in a stream of fluid. *Phil. Trans. R. Soc.* A21A:374–443.
3. Hughes, J. A. 1916. On the cooling of cylinders in a stream of air. *Phil. Mag.* 31:181.
4. Worthington, H. V., and Malone, C. B. 1917. Convection of heat from small wires in water. *J. Franklin Inst.* 184:115.
5. Davis, A. H. 1924. Convective cooling of wires in streams of viscous liquids. *Phil. Mag.* 47:282.
6. Žukauskas, A. A., Makarevičius, V. J., and Šlančiauskas, A. A. 1968. Teplootdacha puchkov trub v poperechnom potoke zhidkosti (Heat transfer in banks of tubes in cross flow of Fluid). Vilnius, Mintis.
7. Žukauskas, A. A. 1972. Heat transfer from tubes in cross-flow. In *Advances in Heat Transfer*, vol. 8, pp. 93–160. New York: Academic Press.
8. Collis, D. S., and Williams, M. J. 1959. Two-dimensional convection from heated wires at low Reynolds numbers. *J. Fluid Mech.* 6(3):357–384.
9. Davis, P. O., and Fischer, M. J. 1964. Heat transfer from electrically heated cylinders. *Proc. R. Soc.* A280:486.
10. Piret, E. L., James, W., and Stacy, M. 1947. Heat transmission from fine wires to water. *Ind. Eng. Chem.* 39:1098–1103.
11. Žukauskas, A. A., and Indriunas, A. A. 1954. Vieleliu šilumos atidavimo skersiniame klampaus skysčio sraute tyrimas. *Fizikos-Technikos Institut Darbai* 1:65–71.
12. Hilpert, R. 1935. Wärmeabgabe von geheizten Drähten und Rohren im Lufstrom. *Forsch. Geb. Ingenieurwes.* 4:215–224.
13. Mikheev, M. A. 1937, 1943. Zavisimost teploobmena ot napravlenya teplovogo potoka The dependence of the heat transfer on the heat flux direction). Izvestia Akademii Nauk SSSR, Otdel Tekhnicheskikh Nauk, no. 3, p. 335. *Zh. Tekhnicheskoi Fiziki* 13(6):311.

14. Kruzhilin, G. N., and Shvab V. A. 1935. Issledovanya α-polya na poverchnosti kruglogo tsilindra, omyvaemogo poperechnym potokom vozdukha, v intervale znachenyi kriterya Reynoldsa ot $21 \cdot 10^3$ do $85 \cdot 10^3$ (A study of the α-field on a cirdular cylinder in cross flow of air for Re from $21 \cdot 10^3$ to $85 \cdot 10^3$). *Zh. Tekhnicheskoi Fiziki* 5(3): 483–488; 5(4):707–710.
15. Kruzhilin, G. N. 1938. Teplootdacha kruglogo tsilindra v poperechnom potoke vozdukha v intervale znachenyi chisla Reynoldsa ot 6000 do 425000 (Heat transfer of a circular cylinder in cross flow of air for Re from 6000 to 425000). *Zh. Tekhnicheskoi Fiziki* 8(2):25–30.
16. Schmidt, E., and Wenner, K. 1941. Wärmeabgabe über den Umfang eines angeblesenen geheizten Zylinders. *Forsch. Geb. Ingenieurwes.* 12:65–73.
17. Kruzhilin, G. N. 1936. Teorya teploperedachi kruglogo tsilindra v poperechnom potoke zhidkosti (A theory of the heat transfer for a circular cylinder in cross flow). *Zh. Tekhnicheskoi Fiziki* 6(5):858–865.
18. Frössling, N. 1940. Verdunstung, Wärmeübergung und Geschwindigkeitsverteilung bei zweidimensionaler und rotationasimmetrischer Grenzschichtströmung. *Lunds Univ. Arssk. N.F. Avd. 2* 36(4):25–35.
19. Eckert, E. R. G. 1942. Die Berechnung des Wärmeüberganges in der laminaren Grenzschicht umströmter Körper. *VDI-Froschungsheft* 416:1–26.
20. Krall, K. M., and Eckert, E. R. G. 1970. Heat transfer to a transverse circular cylinder at low Reynolds number including refraction effects. In *Heat Transfer*, vol. 3, pp. 225–232. Paris-Versailles.
21. Gomelauri, V. I. 1948. K voprosu o teplootdache v potoke kapelnykh zhydkostey (On the heat transfer in a flow of liquid droplets). *Tr. Energeticheskogo Instituta Akademii Nauk Gruzinskoy SSR* 4(4).
22. Žukauskas, A. A. 1955. Teplootdacha tsilindra v poperechnom potoke zhidkosti (Heat transfer of a cylinder in cross flow). *Teploenergetika* 4:38–40.
23. Žukauskas, A. A. 1959. Teploperedacha pri poperechnom omyvanyi tsilindra (Heat transfer of a cylinder in cross flow). In: *Teploperedacha i Teplovoe Modelirovanye*, pp. 201–212. Moscow: Akademya Nauk SSSR.
24. Daujotas, P. M., Žiugžda, J. J., and Žukauskas, A. A. 1973. Teplootdacha tsilindra v poperechnom potoke vody v oblasti kriticheskikh znachenyi chisla Reynoldsa (Heat transfer of a cylinder in cross flow of water in the critical range of the Reynolds number). *Lietuvos TSR Mokslu Akademijos Darbai, Ser. B.* 3(76):99–109.
25. Zdanavičius, G. B., Česna, B. A., Žiugžda, J. J., and Žukauskas, A. A. 1975. Mestnaya teplootdacha poperechno obtekaemogo potokom vozdukha kruglogo tsilindra pri bolshykh znachenyakh Re (Local heat transfer of a circular dylinder in cross flow of air at high values of Re). *Lietuvos TSR Mokslu Akademijos Darbai, Ser. B* 2(87): 109–119.
26. Achenbach, E. 1975. Total and local heat transfer from a smooth circular cylinder in cross-flow at high Reynolds number. *Int. J. Heat Mass Transfer* 10(12):1387–1396.
27. Roshko, A. 1961. Experiments on the flow past circular cylinders at a very high Reynolds number. *J. Fluid Mech.* 10(3):345–356.
28. Tani, I. 1964. Low flows involving bubble separations. In *Progress in Aeronautic Science*, pp. 70–103. Oxford: Pergamon Press.
29. Eygenson, L. S. 1935. K voprosu o vlyanii turbulentnosti na teplootdachu (On the effect of turbulence on the heat transfer). *Izvestia Energeticheskogo Instituta Akademii Nauk SSSR* 3(1–2):29–35.
30. Kirpichev, M. V., and Eygenson, L. S. 1936. Teplovoi turbulimetr (A thermal meter of turbulence). *Izvestia Energeticheskogo Instituta Akademii Nauk SSSR* 4(1):81–89.
31. Giedt, W. 1958. Effect of turbulence level of incident air stream on local heat transfer and skin friction on a cylinder. *J. Aeron. Sci.* 18(11):725–730, 766.

32. Kestin, J., and Wood, R. 1971. Influence of turbulence on mass transfer from cylinders. *Trans. ASME* C93(4):1–8.
33. Kestin, J. 1966. The effect of free-stream turbulence on heat transfer rates. In *Advances in Heat Transfer*, vol. 3, pp. 1–32. New York: Academic Press.
34. Dyban, E. P., Epik, E. Y., and Kozlova, L. G. 1972. Teploobmen i gidrodinamika krugovogo tsilindra, poperechno obtekaemogo turbulizirovannym vozdushnym potokom (Heat transfer and fluid dynamics on a circular cylinder in a turbulent cross flow of air). In *Teplo-i Massoperenos*, vol. I., pp. 222–226. Minsk.
35. Lowery, G. W., and Vachon, R. I. 1975. The effect of turbulence on heat transfer from heated cylinders. *Int. J. Heat Mass Transfer* 18(11):1229–1242.
36. Katinas, V. J., Žiugžda, J. J., Žukauskas, A. A., and Švegžda, S. A. 1974. Vlyanie turbulentnosti nabegayuschego potoka vyazkoi zhidkosti na mestnuyu teplootdachu kruglogo tsilindra (The effect of free stream turbulence of a viscous fluid on the local heat transfer of a circular cylinder). *Lietuvos TSR Mokslu Akademijos Darbai, Ser. B.* 5(84):129–145.
37. Zdanavičius, G. B., Survila, V. J., and Žukauskas, A. A. 1975. Vlyanye stepeni turbulentnosti nabegayuschego potoka vozdukha na mestnuyu teplootdachų kruglogo tsilindra v kriticheskoi oblasti obtekanya (The effects of free stream turbulence of air on the local heat transfer of a circular cylinder in the critical flow regime). *Lietuvos TSR Mokslu Akademijos Darbai, Ser. B* 4(89):119–129.
38. Ilgarubis, V. S., Daujotas, P. M., Žiugžda, J. J., and Žukauskas, A. A. 1977. Teplootdacha tsilindra, poperechno obtekaemogo turbulizirovannym potokom vody v oblasti kriticheskikh znachenyi Re (Heat transfer of cylinder in a turbulent cross flow of water in the critical range of Re). *Lietuvos TSR Mokslu Akademijos Darbai, Ser. B* 3(100):91–103.
39. Akylbaev, Z. S., Isataev, S. I., Krashtalev, P. A., and Masleeva, N. V. 1966. Vlyanie zagromozhdenya potoka na koeffitsient mestnoi teplootdachi odnorodnogo nagretogo tsylindra (The effect of channel blockage on the local heat transfer coefficient of a uniformly heated cylinder). In *Problemy Teploenergetyki i Prikladnoi Teplofiziki*, vol. 3, pp. 179–198. Alma-Ata: Nauka.
40. Epik, E. Y., and Kozlova, L. G. 1973. Vlyanie zagromozhdenya kanala i turbulentnosti potoka na obtekanye tsilindra (The effect of channel blockage and free stream turbulence on the fluid dynamics over a cylinder). *Teplofizika i Teplotekhnika* 25: 55–57.
41. Perkins, H., and Leppert, G. 1964. Local heat transfer coefficients on a uniformly heated cylinder. *Int. J. Heat Mass Transfer* 7(2):143–158.
42. Achenbach, E. 1977. The effect of surface roughness on the heat transfer from a circular cylinder to the cross flow of air. *Int. J. Heat Mass Transfer* 20(4):359–369.
43. Leontyev, A. I., and Ryagin, B. A. 1966. Teploobmen v vikhrevoy oblasti pri poperechnom obtekanyii tsilindra (Heat transfer in the vortical region of a cylinder in cross flow). *Zh. Prikladnoi Matematiki i Tekhnicheskoi Fiziki* 6:111–114.
44. Virk, P. S. 1970. Heat transfer from the rear of a cylinder in transverse flow. *Trans. ASME, Ser. C* 92(1):206–207.
45. Igarashi, T., and Hirata, M. 1974. Heat transfer in separated flows. In *5th Int. Heat Transfer Conf. Tokyo*, 1974, vol. 2, pp. 300–308.
46. Mabuchi, I., Hivada, M., and Kumada, M. 1974. Some experiments, associated with heat transfer mechanism in turbulent separated region of a cylinder. In *5th Int. Heat Transfer Conf. Tokyo*, 1974, vol. 2, pp. 315–319.
47. Kostič, Z. G., and Oka, S. N. 1972. Fluid flow and heat transfer with two cylinders in cross flow. *Int. J. Heat Mass Transfer* 15(2):279–299.
48. Morton, V. T. 1975. The overall convective heat transfer from smooth circular cylinders. In *Advances in Heat Transfer*, vol. 11, pp. 199–264. New York: Academic Press.

49. Gnielinski, V. 1975. Berechnung mittlerer Wärme- und Stoffüberganskoeffizienten an laminar und turbulent überströmten Einzelkörpern mit Hilfe einer einheitlichen Gleichung. *Forsch. Ingenieurwes.* 41(5):145–153.
50. Börner, H. 1965. Über den Wärme- und Stoffübergang an umspülten Einzelkörpern bei Überlagerung von freier und erzwungener Strömung. *VDI-Forschungsheft*, Düsseldorf, vol. 512.
51. Churchill, S. W., and Bernstein, M. 1977. A correlating equation for forced convection from gases and liquids to a circular cylinder in cross-flow. *Trans. ASME Heat Transfer, Ser. C* 99(2):300–306.
52. Isachenko, V. P., Osipova, V. A., and Sukhomel, A. S. 1975. *Teplootdacha* (*Heat Transfer*). Moscow: Energya.
53. Schlichting, H. 1960. *Boundary-Layer Theory.* New York: McGraw-Hill.
54. Konstantinov, N. I., and Dragnysh, G. L. 1955. K voprosu ob izmerenii napryazhenya na poverkhnosti (On the measurement of shear stress). *Tr. Leningradskogo Politekhnicheskogo Instituta* No. 176, pp. 191–201.
55. Walz, A. 1966. *Strömungs- und Temperaturgrenzschichten.* Karlsruhu: Verlag G. Braun.
56. Eckert, E. R. G., and Soehngen, E. 1950. Distribution of heat-transfer coefficients around circular cylinder in crossflow at Reynolds numbers from 20 at 500. *Trans. ASME* 74:343–350.
57. Lin, C. J., Pepper, D. W., and Lee, S. C. 1976. Turbulent heat transfer at low Reynolds number. *AIAA J.* 14(7):581–590.
58. Traci, R. M., and Wilcox, D. C. 1975. Free stream turbulence effects on stagnation point heat transfer. *AIAA J.* 13(7):890–896.
59. Smith, M. C., and Kuethe, A. M. 1966. Effects of turbulence on laminar skin friction and heat transfer. *Phys. Fluids* 9(12):2337–2344.
60. Kochin, N. E., Kibel, I. A., and Roze, N. V. 1963. *Teoreticheskaya Gidromekhanika (Theoretical Hydromechanics)*, part I. Moscow: GIFML.
61. Borisenko, A. I. 1953. Teorya gidrodinamicheskoy reshetki krugovykh tsilindrov (A theory of fluid dynamics on a grid of circular cylinders). *Tr. Laboratorii Akademii Nauk Ukrainskoi SSR.* Kiev: Akademia Nauk Ukrainskoi SSR.
62. Petukhov, B. S. 1952. *Opytnoe Izuchenye Protsessov Teploperedachi.* Moscow-Leningrad: Gasoenergoizdat.
63. Dyban, E. P., and Epik, E. Y. 1976. Ispolzovanye statisticheskikh kharakteristik turbulentnosti v raschetakh konvektivnogo teploobmena (The application of statistical parameters of turbulence in the heat transfer predictions). In *Teplomassoobmen-5*, vol. I, part I, pp. 25–34. Minsk.
64. McDonald, H., and Kreskovsky, J. P. 1953. Effect of free stream turbulence on the turbulent boundary layer. United Aircraft Research Laboratories, East Hartford, Connecticut.
65. Cebesi, T. 1970. Behavior of turbulent flow near a porous wall with pressure gradient. *AIAA J.* 8:2152–2160.
66. Cebesi, T., and Mosinskis, G. J. 1971. Computation of incompressible turbulent boundary layer at low Reynolds numbers. *AIAA J.* 9:1632–1641.
67. Cebesi, T., and Smith, A. M. O. 1974. *Analysis of Turbulent Boundary Layer.* New York: Academic Press.
68. Bradshaw, P. 1969. The analogy between streamline curvature and buoyancy in turbulent shear flow. *J. Fluid Mech.* 36:177–198.
69. Cebesi, T. 1968. Curvature and transition effects in turbulent boundary layer. *AIAA J.* 9:1091–1100.
70. Chen, K. K., and Thyson, N. A. 1971. Extension of Emmon's spot theory to flows on blunt bodies. *AIAA J.* 5:821–828.
71. Patankar, S. V., and Spalding, D. B. 1970. Heat and Mass Transfer in Boundary Layer. London.

72. Holstein, H., and Bohlen, R. 1940. Lilienthal: Bericht, pp. 105–106.
73. Squire, H. B. 1942. Heat transfer calculation for aerofoils. ARC RM, no. 1986.
74. Katinas, V. J., Švegžda, S. A., Žiugžda, J. J., and Žukauskas, A. A. 1975. Obtekanye lobovoi chasti kruglogo tsilindra turbulizirovannymi potokami vyazkoi zhidkosti i ee teplootdacha (Fluid dynamics and heat transfer over the front part of a circular cylinder in turbulized viscous fluid flows). *Lietuvos TSR Mokslu Akademijos Darbai, Ser. B* 1(86):103–114.
75. Loitsyanskii, L. G. 1973. *Mekhanika Zhidkosti i Gaza (Fluid Mechanics).* Moscow: Nauka.
76. Michel, R. 1951. Etude de la transition sur les profiles d'aile, establissement d'un critère de determination de point de transition et calcul de la trainée de profile incompressible. ONERA Report 1/1578A.
77. Cebesi, T., Mosinskis, G. J., and Smith, A. M. O. 1972. Calculation of viscous drag and turbulent boundary-layer separation on two-dimensional and axisymmetric bodies in incompressible flows. Report N MDC J0973-01. Long Beach, California: Douglas Aircraft Co., N 71-25868.
78. Coles, D. 1962. The turbulent boundary layer in a compressible fluid. Report R-403-PR. Santa Monica, California: Rand Corp., AD 285-651.
79. Žukauskas, A. A., and Šlančiauskas, A. A. 1973. *Teplootdacha v Turbulentnom Potoke Zhidkosti (Heat Transfer in Turbulent Flow of Fluid).* Vilnius: Mintis.
80. Karyakin, Y. E. amd Sharov, V. T. 1974. Konechno-raznostnyi metod rascheta turbulentnogo pogranichnogo sloya neszhimaemoi zhidkosti (A finite-difference technique of determining the turbulent boundary layer of an incompressible liquid). *Inzhenerno-Fizicheskii Zh.* 26(2):191–198.
81. Jones, W. P., and Launder, B. E. 1973. The calculations of low Reynolds number phenomena with a two-equation model of turbulence. *J. Heat Mass Transfer* 16(6): 1119–1130.
82. Hiemenz, K. 1911. Die Grenzschicht an einem in den gleichformigen Flüssigkeitsstrom eingetauschten geraden Kreiszylinder. Drs. Göttingen.
83. Goldstein, S. 1965. *Sovremennoye Sostoyanye Gidroaerodinamiki Vyazkoi Zhidkosti (Modern Developments in Fluid Dynamics)*, vol. 2, Dover: New York.
84. Gerrard, J. 1966. The three-dimensional structure of the wake of a circular cylinder. *J. Fluid Mech.* 25(1):60–81.
85. Kozlova, L. G., and Epik, E. Y. 1969. K voprosu o teploobmene v lobovoi tochke poperechno obtekaemogo tsilindra (On the heat transfer in the front stagnation point of a cylinder in cross flow). *Voprosy Tekhnicheskoi Teplofiziki* 2:69–71.
86. Roshko, A., and Fishdon, U. 1969. O roli perekhoda v blizhnem slede (On the role of transition in the near wake). *Mekhanika* 6(118):50–58.
87. Richardson, E. G. 1961. *Dinamika Realnych Zhidkostei (Dynamics of Real Fluids).* London: Edward Arnold.
88. Drescher, H. 1956. Messung von zeitlich veränderten Drucken. *Z. F. Flugwis.* 4:53–59.
89. Ferguson, N., and Parkinson, G. V. 1967. Surface and wake flow phenomena of the vortex–excited oscillation of a circular cylinder. *J. Eng. Ind. Ser. B* 4:260–269.
90. Heine, W. 1964. On the investigation on vortex excited pressure fluctuations. M. Sc. thesis, University of British Columbia.
91. Švegžda, S. A., Marr, Y. N., Žiugžda, J. J., and Žukauskas, A. A. 1977. Nestatsionarnost techenia okolo tsilindra, poperechno obtekaemogo potokom vozdukha. (Instability of flow over a cylinder in cross flow of air). *Lietuvos TSR Mokslu Akademijos Darbai, Ser. B* 6(103):73–78.
92. Epik, E. Y., and Kozlova, L. G. 1973. Lokalnyi teploobmen poperechno obtekaemogo tsilindra v turbulizirovannom potoke (Local heat transfer of a cylinder in cross flow of a turbulized fluid). *Voprosy Tekhnicheskoi Teplofiziki* 4:44–48.

93. Bearman, P. 1967. On vortex street wakes. *J. Fluid Mech.* 28:321–331.

94. Devnin, S. I. 1967. *Aerogidrodinamicheskii Raschet Plokhoobtekaemykh Sudovykh Konstruktsyi (Fluid-Dynamical Predictions for Non-Slender Constructions of Ships).* Leningrad: Sudostroyenye.

95. Roshko, A. 1955. On the development of turbulent from vortex streets. (In NASA Report 1191, 1954.) *Mekhanika (sb. Perevodov)* 5(89):51–70.

96. Akylbaev, Z. S., Isataev, S. I., and Polzik, V. V. 1972. Sryv vikhrei s poverkhnosti plokhoobtekaemykh tel i ego vlyanye na teploobmen (Vortex shedding from non-slender bodies and its effect on the heat transfer). In *Teplo-i Massoperenos*, vol. 1, part I, pp. 291–295. Minsk.

97. Blasius, H. 1908. Grenzschichten in Flüssigkeiten mit kleiner Reibung. *Z. Math. Phys.* 56:1–37.

98. Howarth, L. 1935. On the calculation of steady flow in the boundary layer near the surface of a cylinder in a stream. ARC RM, no. 1632.

99. Schönauer, W. 1964. Ein Differenzverfahren zur Lösung der Grenzschichtgleichung für stationare laminare, incompressible Strömung. *Ing. Arch.* 33:183–189.

100. Ruseckas, T. B., Žukauskas, A. A., and Žiugžda, J. J. 1979. Vlyanye turbulentnosti nabegayuschego potoka na mestnye kharakteristiki soprotivlenya poperechnoobtekaemykh krivolineynykh tel (The effect of free stream turbulence on the local parameters of resistance of curvilinear bodies in cross flow.) Deposited in the Lithuanian Institute of Technical Information, no. 447-79.

101. Žukauskas, A. A., Vaitiekunas, P. P., and Žiugžda, J. J. 1979. Konechnoraznostnyi metod opredelenya teplootdachi lobovoi chasti poperechno obtekaemogo kruglogo tsilindra s uchetom vlyanya turbulentnosti potoka (A finite-difference determination of the heat transfer of the front part of a cylinder in cross flow with the effects of free stream turbulence). Deposited in the Lithuanian Institute of Technical Information, no. 446–79.

102. Žukauskas, A. A., Ruseckas, T. B., and Žiugžda, J. J. 1979. Issledovanye vlyanya turbulentnosti na integralnye kharakteristiki soprotivlenya poperechno obtekaemykh tsilindricheskikh tel (A study of the effect of turbulence on the integral resistance of cylinders in cross flow). Deposited in the Lithuanian Institute of Technical Information, no. 448-79.

103. Son, J. S., and Hanratty, T. J. 1969. Velocity gradients at the wall for flow around a cylinder at Reynolds numbers from 5×10^3 to 10^5. *J. Fluid Mech.* 35:353–368.

104. Krall, K. M., and Eckert, E. R. 1961. On heat transfer in stagnation region of circular cylinders. *J. Fluid Mech.* 9(2):125–132.

105. Kayalar, L. 1969. Experimentale und theoretische Untersuchungen über den Einfluss des Turbulenzgrades auf den Wärmeübergang in der Umgebung des Staupunkts eines Kreiszylinders. *Forsch. Ingenieurwes.* 35(5):158–167.

106. Kestin, J., and Wood, R. T. 1970. The mechanism which causes free stream turbulence to enhance stagnation-line heat and mass transfer. In *Heat Transfer 1970, Fourth Int. Heat Transfer Conf.*, FC 2.7, Paris, vol. 3, pp. 181–188.

107. Čolak-Antič, P. 1971. Visuelle Untersuchungen von Längswirbeln im Staupunktgebiet eines Kreiszylinders bei turbulenter Anströmung. Bericht über die DGLR Fachausschuss-Sitzung. In *Laminare und Turbulente Grezschichten*, 7–8 June 1971, Göttingen.

108. Hassler, H. 1971. Hitzdrahtmessungen von längratigen Instabilitätserscheinungen im Staupunktgebiet eines Kreiszylinders in turbulent Anströmung. In *Laminare und Turbulente Grezschichten*, 7–8 June, 1971, Göttingen.

109. Burno, E., Diep, G. B., and Kestin, J. 1966. Sur un nouveau type de tourbillons longitudinaux ans l'écoulement autour d'un cylindre. *Co. R. Acad. Sci. Paris* 263:742.

110. Dyban, E. P., Epik, E. Y., and Kozlova, L. G. 1975. Sovmestnoe vlyanye stepeni, prodolnogo masshtaba turbulentnosti i uskorennosti vozdushnogo potoka na teploobmen kruglogo tsilindra (A combined effect of turbulence, its longitdinal scale and flow acceleration on the heat transfer of a circular cylinder in air). In *Teploobmen, 1974.* Moscow: Sovetskye Issledovanya, Moscow, Nauka.
111. Katinas, V. J., Žiugžda, J. J., and Žukauskas, A. A. 1970. Teplootdacha krivolineynykh tel pri poperechnom ikh obtekanii viazkoi zhidkosti (Heat transfer of curvilinear bodies in cross flow of viscous fluids). *Lietuvos TSR Mokslu Akademijos Darbai, Ser. B* 4(63):209–233.
112. Merk, H. J. 1959. Rapid calculations for boundary layer transfer using wedge solutions and asymptotic expansions. *J. Fluid Mech.* 5(3):460–480.
113. Makarevičius, V. J., Žiugžda, J. J., and Žukauskas, A. A. 1962. K voprosu rascheta teplootdachi krivolineynykh poverkhnostei pri laminarnom pogranichnom sloe (Heat transfer predictions for laminar boundary layers on curvilinear bodies). *Lietuvos TSR Mokslu Akademijos Darbai, Ser. B* 3(30):191–202.
114. Junkhan, G. H., and Serovy, G. H. 1967. Effects of free-stream turbulence and pressure gradient of flat plate boundary layer velocity profiles and heat transfer. *Trans. ASME, J. Heat Transfer,* Ser. C, 89(2):58–68.
115. Dyban, E. P., Epik, E. Y., and Suprun, T. T. 1976. Kharakteristiki laminarnogo pogranichnogo sloya pri povyshennoi turbulentnosti nabegayuschego potoka (The parameters of the laminar boundary layer for high free stream turbulence). *Teplofizika i Teplotekhnika* 30:86–95.
116. Kažimekas, P. 1978. Vlyanye turbulentnosti vneshnego potoka zhidkosti na teplootdachu plastiny (The effect of free stream turbulence on the heat transfer of a plate). *Lietuvos TSR Mokslu Akademijos Darbai, Ser. B* 6(109):47–52.
117. Daujotas, P. M., Žiugžda, J. J., and Žukauskas, A. A. 1975. Teplootdacha razlichnykh oblastey poperechno obtekaemogo potokom vody tsilindra pri kriticheskikh znachenyakh Re (Heat transfer in different regions of a cylinder in cross flow of water at critical values of Re). *Lietuvos TSR Mokslu Akademijos Darbai, Ser. B* 1(86):95–101.
118. Žukauskas, A. A., Ilgarubis, V. S., and Žiugžda, J. J. 1977. Vlyanye zagromozhdenya kanala na mestnuyu teplootdachy poperechno obtekaemogo tsilindra pri kriticheskikh Re (The effect of channel blockage on the local heat transfer of a cylinder in cross flow at critical Re). Deposited in the Soviet Institute of Technical Information, Academy of Sciences, USSR, N 2737-77, 16, p. 119.
119. Richardson, P. D. 1963. Heat and mass transfer in turbulent separated flows. *Chem. Eng. Sci.* 18:149–155.
120. Spalding, D. B. 1967. Heat transfer from turbulent separated flows. *J. Fluid Mech.* 27(1):97–109.
121. Žukauskas, A. A., and Žiugžda, J. J. 1969. *Teplootdacha v Laminarnom Potoke Zhidkosti (Heat Transfer in Laminar Flow of Fluid).* Vilnius, Mintis.
122. Makarevičius, V. J. 1978. *Teploobmen pri Fiziko-Khimicheskikh Izmenenyakh (Heat Transfer in the Presence of Physical-Chemical Factors).* Vilnius, Mokslas.
123. Kuznetsov, N. V. 1958. *Rabochye Prostessy i Voprosy Usovershenstvovanya Konvektivnykh Poverkhnostey Kotelnikh Agregatov (Working Processes and Improvement of Convective Surfaces in Boilers).* Moscow: Gosenergoizdat.
124. Mikheev, M. A. 1952. Teplootdacha pri turbulentnom dvizhenii zhidkosti v trubkakh (Heat transfer in turbulent pipe flows). *Izvestia Akademii Nauk SSSR* (Otdel Tekhnicheskikh Nauk) 10:128–139.
125. Gregorig, R. 1970. Verallgemeinter Ausdruck für den Einfluss temperaturabhängiger Stoffwerte auf den turbulenten Wärmeübergang. *Wärme- und Stoffübertragung* 3:26–40.
126. Mikheev, M. A., and Mikheeva, I. M. 1973. *Osnovy Teplootdachi (Fundamentals of Heat Transfer).* Moscow: Energya.

127. Hatton, A. P., James, D. D., and Swise, H. W. 1970. Combined forced and natural convection with low-speed air flow over horizontal cylinders. *J. Fluid Mech.* 42: 17–31.
128. Van der Hegge Zijnen, B. G. 1956. Modifical correlation formulae for the heat transfer by forced convection from horizontal cylinders. *Appl. Sci. Res., Sect. A* 6:129–140.
129. Isataev, S. I., Akylbaev, Z. S., Masleeva, N. V., and Polzik, V. V. 1968. Teplootdacha tsilindra i sfery v kanale so znachitelnym zagromozhdenyen (Heat transfer of a cylinder and a sphere in a channel for high blockage factors). In *Teplo-i Massoperenos*, vol. 1, pp. 320–329. Moscow: Energya.
130. Gimbutis, G. I., and Šhapola, V. I. 1972. K voprosu teplootdachi pri poperechnom obtekanii tsilindra vozdukhom (On the heat transfer of a cylinder in cross flow of air). In *Mekhanika*, pp. 226–227. Kaunas.
131. Žukauskas, A. A., Simanavičius, V. S., Daujotas, P. P., and Žiugžda, J. J. 1978. Teplootdacha sherokkovatogo tsilindra v poperechnom potoke vody pri kriticheskikh znachenyakh Re (Heat transfer of a rough-surface cylinder in cross flow of water at critical values of Re). *Lietuvos TSR Mokslu Akademijos Darbai, Ser. B* 2(105):83–92.
132. Žukauskas, A. A., Žiugžda, J., and Daujotas, P. 1978. Effects of turbulence on the heat transfer of a rough-surface cylinder in cross flow in the critical range of Re. In *6th Int. Heat Transfer Conf., Toronto*, vol. 4, pp. 231–236.
133. Shitnikov, V. K. 1961. K voprosu o vylyanii formy na protsess vneshnego teploobmena lri vinuzhdennoi konvektsii (The effect of geometry on the external forced-convection heat transfer). *Inzhenerno-Fizicheskii Zh.* 6:78–82.
134. Suleimanova, L. L., and Gurenkova, T. V. 1976. Zavisimost radiatsionnoi sostavlyayuschei koeffitsienta teploprovodnosti zhidkostei ot temperatury (The effect of temperature on the radiation component of the heat conduction coefficient in fluids). In *Teplo-i Massoobmen v Khimicheskoi Tekhnologii*, vol. 4, pp. 21–24. Kazan.
135. Gorshenina, T. N., and Gurenkova, T. V. 1976. Radiatsionnaya sostavlyayuschaya teploprovodnosti geksana i nekotorykh ego proizvodnykh (Radiation component in the heat conduction of hexane and some of its derivatives). In *Teplo-i Massoobmen v Khimicheskoi Tekhnologii*, vol. 4, pp. 24–27, Kazan.
136. Vargaftik, N. B. 1972. *Spravochnik po Teplofizicheskim Svoistvam Gazov i Zhidkostey (Data Book of Thermal Properties for Fluids).* Moscow: Nauka.
137. Žukauskas, A. A., and Leizerson, A. N. 1969. Teplootdacha pryamougolnogo sterzhnya v potoke zhidkosti (Heat transfer of a rectangular rod in a liquid flow). *Lietuvos TSR Mokslu Akademijos Darbai, Ser. B* 4(51):95–109.

NOMENCLATURE

a	thermal diffusivity, m/s
c_D	hydraulic drag coefficient
c_w	pressure drag coefficient
c_p	specific heat capacity, J/kg·K
c_f	skin friction coefficient
d	cylinder diameter, m
d_1	major axis of the elliptic cylinder, m
d_2	minor axis of the elliptic cylinder, m
F	surface area, m^2
f	frequency, Hz
H	height of the test section, m
I_x	integral Euler time scale
k	exponent on Tu
k	height of roughness element, m
k_q	blockage factor, d/H
l	length of cylinder
l	longitudinal distance from the front stagnation point, m
L	macroscale length, m
m, n	exponents of Reynolds and Prandtl numbers
p	pressure, N/m^2
p	exponent of Pr_f/Pr_w
p_∞	static pressure of free stream, N/m^2

p_φ local pressure on cylinder surface, N/m^2
$\bar{p}$ pressure coefficient, $2(p_\varphi - p_\infty)/\rho U_\infty^2$
Δp pressure drop, N/m^2
$p_{180°}$ base pressure, $2(p_{180°} - p_\infty)/\rho U_\infty^2$
Q heat rate, W
q_w heat flux density, W/m^2
r radius, m
r_1, r_2 inner and outer radius of the cylinder, m
$R_{x(\tau)}$ Euler time correlation coefficient
S surface curvature
Tu turbulence intensity, %
t, T temperature, °C
Δt temperature difference, °C
U_∞ free-stream velocity, m/s
U_φ maximum velocity outside the boundary layer, m/s
U velocity in the least free cross section, m/s
u streamwise mean (bulk) velocity, m/s
u' longitudinal component of the velocity fluctuation, m/s
$\bar{u}'$ mean-square value of longitudinal velocity fluctuation, m/s
u_* friction velocity, m/s
v mean (average) velocity, normal to the surface, m/s
x, y Cartesian coordinates
x distance from the grid to the front stagnation point, m
α_x, α local and average coefficient of heat transfer, respectively, $W/m^2 \cdot K$
γ intermittency factor
δ thickness of the hydrodynamical boundary layer, m
δ_T thickness of the thermal boundary layer, m
δ^* displacement thickness, m
δ^{**} momentum thickness, m
κ von Karman constant
λ thermal conductivity, $W/m \cdot K$
Λ spatial macroscale, m
μ dynamic viscosity, $N \cdot s/m^2$
ν kinematic viscosity, m^2/s
ρ density, kg/m^3
τ_w shear stress, N/m^2
$\bar{\tau}$ dimensionless shear stress, $(2\tau_w/\rho u_\infty)\sqrt{Re}$
φ angular distance from the front stagnation point, degrees
Nu Nusselt number
Pr Prandtl number
Pr_t turbulent Prandtl number
Re Reynolds number

Ri	Richardson number
Sr	Strouhal number

Subscripts

0	entrance cross section
f, ∞	free stream (main flow)
w	wall
t	turbulent flow
s	separation condition
x, φ	local condition

INDEX

Achenbach, E., 5, 48, 63, 117, 147
Akylbaev, Zh. S., 31, 120, 144
Anemometer:
thermal (hot-wire), 23

Bearman, P., 56
Bernoulli equation, 20, 46, 71
Bernstein, M., 139
Blasius, H., 58
Blockage factor, 5, 48, 56, 77, 119, 145, 162
effect of, 27, 31, 47, 126
Börner, H., 7
Boundary layer:
equation, 4, 27, 104
on front part, 83
laminar, 1, 5, (illus.) 34
on rear part, 155
reattachment of, 155
separation of, 30, 36, 39, 48, 113
thickness of, 37, 81
transitional, 37
(*See also* Transition of boundary layer)
turbulent, 38
separation of, 36, 159
thermal, 111
Bradshaw, 39

Cebesi, T., 27, 37
Churchill, S. W., 7, 138
Collis, D. S., 8, 135
Conductivity:
thermal, 33
turbulent, 104
Convection, 134
combined, 134, 135
free, 134
Cylinder:
circular, 58, 79, 154
with different profiles, 150
elliptic, 45, 57, 63, 71, 89, (table) 182
heated, (illus.) 16
rough-surface, 6, 118
in series, 151
smooth-surface, 13

Damping factor, 38
Daujotas, P. M., 5, 116
Davis, A. H., 136
Devnin, S. I., 57, 75
Drag:
friction, 74, 156, 157
hydraulic, 56, 73, 156, (table) 190
pressure, 74, 77, 156
Dyban, E. P., 86, 106

Eckert, E. E. G., 27, 82, 87, 98
Eigenson, L. S., 5
Epik, E. Ya., 140
Equivalent diameter, 150
Euler equation, 80
Exponent (*see* Power index)

Flow regime, 162
  critical, 5, 47, 60, 110, (illus.) 115, 121
  subcritical, 48, 104, 137
  supercritical, 134, 138, 159
Fluctuation, roughness-generated, 148
Friction:
  coefficient, 6
  effect of, 73
Frössling, N., 71, 98

Giedt, W., 112
Gimbutis, G. I., 145
Gnielinski, V., 7
Goldstein, S., 40
Gomelauri, V. I., 4
Gregorig, R., 134

Hanratty, T. J., 49
Hatton, A. P., 135
Heat flux, 33
  constant, 16
  direction of, 4, 63
  on surface, 19
Heat transfer, 4
  average, 5, 7, 14, 18, 140, 150, 165, (table) 173
  fluid-wall, 130
  in front stagnation point, 79, 85, 88
  in laminar boundary layer, 102
  local, 14, 97, 110, (illus.) 115, 153, (table) 168
  in rear part, 107, 109, 138
  at rear stagnation point, 116
  on rough-surface, 146
  from tubes, 135
  tube-to-fluid, 3, 4
    (*See also* wall-to-fluid)
  from wires, 2, 135
  wall-to-fluid, 130
    (*See also* tube-to-fluid)
Heat transfer coefficient, 2, 3
  average, 3, 14, 18, 116, 129, 137
  for cylinders, 129, 136
  for front half, 106
  at front stagnation point, 89, 93
  local, 3, 14, 98, 101, 109
  at rear stagnation point, 126
  in rear zone, 103
  for wires, 136
    (*See also* Heat transfer)
Hiemenz, K., 28, 58, 71
Hiemenz relation, 28, 31
Hilpert, R., 3, 135
Holstein, H., 29
Howarth, L., 58
Hydraulic drag coefficient, 158
  (*See also* Drag)

Ilgarubis, V. S., 89
Integral equation, 4
Isachenko, V. P., 7

Jones, W. P., 31

Katinas, V. J., 86, 102
Kayalar, L., 83
Kestin, J., 83, 86
Kirpichev, M. V., 5
Krall, K. M., 83, 98
Kochin, N. E., 28
Konstantinov, N. I., 21
Kruzhilin, G. N., 3, 30, 111
Kuethe, A. M., 33

Launder, B. E., 31
Leontyev, A. I., 7
Leppert, G., 123, 145
Lin, C. J., 27
Loicianskij, L. G., 31
Loop, 14
  (*See also* Rig)

Makarevičius, V. J., 2, 131
McDonald, H., 33
Merk, H. J., 98
Michel, R., 37
Mikheev, M. A., 3, 133
Momentum equation, 57
Morton, V. T., 7

Navier-Stokes equation, 27

Oka, S. N., 7
Osipova, V. A., 7

Patankar, S. V., 31, 41
Perkins, H., 123, 145
Petukhov, B. S., 73
Physical properties, 4
 effects of, 4, 106, 132, 160, (illus.) 161
Point:
 of separation, 4, (illus.) 32, 48, 57, 111, 154
 of stagnation:
  front, 27, 30, 46, 51, 59
  rear, 47, 60
 of transition, 51, 114
Power index *n*, (illus.) 131
Prandtl number, 4, 8, 129, 132
Pressure:
 coefficient of, 6
 distribution of, 20, 45, 58, 66, 154, (table) 183
 drop, 5
 gradient, 48, 60
Profile:
 temperature, 35
 velocity, 34

Region:
 separation, 132
  (*See also* Boundary layer)
 separation bubble, (illus.) 55, 113
  (*See also* Separation bubble)
 transitional, (illus.) 55, 113
 turbulent, 132
Reynolds analogy, 86
Reynolds number, 8, 47, 129, 163
 critical, 50
 high, 4, 138
 low, 1, 100, 134
 subcritical, 4
Richardson, P. D., 126
Richardson number, 40
Rig, 10
 for air, 10, (illus.) 12
 of transformer oil, (illus.) 10
 of water, 10, (illus.) 11
Roschko, A., 33, 47, 56, 113
Ruseckas, T. B., 58

Scale:
 integral, 25
 of turbulence (*see* Turbulence)
Schlichting, H., 7, 31
Schmidt, E., 3, 111
Schönauer, W., 58
Separation bubble, 5, 49, 55, 62, 113, 121
Shear stress, 21, (illus.) 34, 57, 64
 distribution, 58, (table) 186
 (*See also* Skin friction)
Schvab, V. A., 3
Skin friction:
 measurement, 26
 probe, 21
 (*See also* Shear stress)
Šlančiauskas, A. A., 2, 31
Smith, A. M., 27, 37
Smith, M. C., 33
Spalding, D. B., 31, 41, 127
Splitter plate, 103
Strouhal, 54, 56
Structure:
 cellular, 83, 103, 165
 three-dimensional, 47
 turbulent, 6
 vortex, 2
Sukomel, A. S., 7
Surface curvature, 39
Surface roughness, 6, 117, 163
 effect of, 17, 95, 118, 160

Tani, I., 113
Temperature, 17
 bulk, 4
 difference, 4
 drop, 18
 free stream, 4
Theory of similarity, 3, 86
Traci, R. M., 27, 94
Transition of boundary layer, 36
 laminar-turbulent, 49, 111, 122, 155, 159
 (*See also* Boundary layer)
Tube:
 circular, 7, 8
 elliptic, 7, 8
 noncircular, 149
 prismatic, 7, 8
 (*See also* Cylinder)

Turbulence:
free stream, 5, 8, 14, 23, 60
effect of, 27, (illus.) 34, 47, 66, 91, 101, 114, 140, 158
scale, 11, 14, 25, 93, 114

Vaitiekunas, P. P., 73
Velocity:
distribution, 28, 35, 73, 154
outside boundary layer, 20, 31, 156
reference, 145
Virk, P. S., 127
Viscosity, turbulent, 33, 104
Vortex shedding, 6, 47, 51, 103, 114, 151 154
Vortex street, von Karman, 47
Vortex structure, 2, 3
(*See also* Structure)

Wake, 2, 48
Wenner, K., 3, 111
Wilcox, D. C., 94
Wood, R. T., 83

Ždanavičius, G. B., 5
Žiugžda, J. J., 5, 86, 98, 146
Žukauskas, A. A., 2, 4, 31, 89, 121